Fermentation Microbiology and Biotechnology

Fermentation Microbiology and Biotechnology

Adrian Vasquez
Joseph Cook

Kruger Brentt
Publishers
2025

Kruger Brentt Publishers UK. LTD.
Company Number 9728962

Regd. Office: 68 St Margarets Road, Edgware, Middlesex HA8 9UU

ISBN: 978-1-78715-439-1

For information on all our publications visit our website at http://krugerbrentt.com/

PREFACE

Fermentation microbiology and biotechnology is a multidisciplinary field that merges principles of microbiology, biochemistry, and engineering to study and optimize the biochemical processes in which microorganisms, like bacteria, yeast, and fungi, convert organic compounds into valuable products. Fermentation, an ancient practice used in food preservation and alcohol production, has evolved to serve broader applications, from pharmaceuticals and biofuels to enzymes and bioplastics. Microbial fermentation, a natural metabolic process, primarily occurs in anaerobic (oxygen-free) conditions and enables the breakdown of complex molecules such as carbohydrates into simpler substances, including gases, acids, or alcohols. As microorganisms metabolize these substrates, they produce products that can enhance flavors, preserve food, or create entirely new products that are not naturally available in the raw materials themselves.

The field of fermentation microbiology focuses on understanding the role of different microorganisms in fermentation processes, optimizing conditions to enhance product yield, and developing methods to control microbial activity. Microorganisms such as Saccharomyces cerevisiae (brewer's yeast), Lactobacillus (lactic acid bacteria), and Clostridium (butyric acid bacteria) are crucial to these processes. Each microorganism has unique properties, fermentation pathways, and environmental requirements, which makes understanding and selecting the right microbe essential for achieving desired outcomes. For instance, the selection of Saccharomyces cerevisiae is critical in alcohol fermentation due to its ability to rapidly convert sugars into ethanol and carbon dioxide, while lactic acid bacteria play a significant role in yogurt and sauerkraut production by producing lactic acid, which enhances flavor and acts as a natural preservative.

Biotechnology extends these fermentation principles by employing advanced techniques to improve efficiency, stability, and scalability of microbial fermentation processes. Techniques such as genetic engineering, metabolic pathway optimization, and synthetic biology are used to modify microorganisms at the genetic level, enabling them to produce higher yields, withstand stressful industrial environments, or synthesize novel compounds that are challenging to produce through traditional methods. Metabolic engineering, for example, allows scientists to manipulate specific pathways within a microorganism to increase the production of a target compound or reduce the formation of by-products. This is critical in the production of biofuels, where enhanced metabolic pathways in yeast or bacteria are used to produce bioethanol or biodiesel on a large scale. Additionally, advancements in bioreactor design, automation, and process control have allowed for more efficient and consistent fermentation outcomes, with precise control over pH, temperature, oxygen levels, and nutrient supply.

The applications of fermentation microbiology and biotechnology are extensive. In the pharmaceutical industry, fermentation is used to produce antibiotics, insulin, and vaccines, which are essential for human health. In the food industry, microbial fermentation produces various fermented foods like cheese, yogurt, soy sauce, and kombucha, which are valued for their nutritional benefits and unique flavors. Moreover, as sustainability becomes a global priority, fermentation biotechnology has gained attention as a potential solution for environmentally friendly biofuel production, bioremediation, and biodegradable material synthesis. The development of bioplastics, produced by microbial fermentation, presents an eco-friendly alternative to conventional plastics, contributing to waste reduction and environmental preservation.

Adrian Vasquez

Joseph Cook

CONTENTS

Section V: Fermentation Microbiology and Biotechnology: Tools, Monitoring, and Control of Fermentation Processes

SECTION I:

FERMENTATION MICROBIOLOGY AND BIOTECHNOLOGY: CENTRAL AND MODERN CONCEPTS

CHAPTER 1

FERMENTATION MICROBIOLOGY AND BIOTECHNOLOGY: AN HISTORICAL PERSPECTIVE

Fermentation is a process that has been a cornerstone of human civilization for thousands of years, used in diverse applications from food production to medicine. This process involves the conversion of carbohydrates to alcohol or organic acids through the activity of microorganisms, primarily bacteria and yeast. Over time, fermentation has evolved from a traditional technique to a sophisticated biotechnological process with significant industrial applications.

1.1 EARLY BEGINNINGS OF FERMENTATION

The origins of fermentation are deeply rooted in ancient civilizations. Archaeological evidence suggests that the Egyptians, Sumerians, and other ancient cultures utilized fermentation processes as far back as 7000 BCE. For instance, the production of beer in ancient Egypt and Mesopotamia marked some of the earliest forms of fermentation.

These early processes were largely accidental, with ancient brewers unaware of the scientific mechanisms behind the transformation of ingredients. Wine, another fermented beverage, also has a long history, with evidence of winemaking dating back to around 6000 BCE in the Caucasus region. Similarly, fermented foods such as yogurt, cheese, and leavened bread were staples in many cultures, underscoring the global significance of fermentation long before the advent of modern science. In the ancient world, fermentation was viewed as a natural, almost magical process. The mystical transformation of fruits, grains, and other foods into new, enhanced products was often associated with religious and cultural practices.

Though people lacked a scientific understanding of fermentation, they intuitively recognized that certain conditions—such as warmth, moisture, and the presence of particular ingredients—were necessary for successful fermentation.

1.2 DISCOVERY OF MICROORGANISMS AND FERMENTATION SCIENCE

The scientific study of fermentation took a monumental step forward with the development of microbiology in the 17th century. The invention of the microscope by Antonie van Leeuwenhoek enabled scientists to observe microorganisms for the first time, although the role of these microbes in fermentation would not be understood until much later. It was not until the 19th century, with the pioneering work of Louis Pasteur, that fermentation was definitively linked to the action of living organisms.

Pasteur's experiments in the 1850s and 1860s demonstrated that fermentation was driven by microorganisms, specifically yeast in the case of alcohol production. This groundbreaking discovery not only challenged the prevailing theory of spontaneous generation but also laid the foundation for the science of microbiology.

Pasteur's research led to the development of "pasteurization," a process that would become essential in ensuring the safety and longevity of fermented food and beverages.

By heating substances to eliminate pathogenic microbes while preserving beneficial ones, pasteurization made it possible to control and stabilize fermentation processes. Pasteur's work also opened the door for more in-depth studies into the metabolic processes of microorganisms and their role in different types of fermentation, such as lactic acid fermentation in yogurt and sauerkraut.

1.3 THE INDUSTRIALIZATION OF FERMENTATION

With the Industrial Revolution came a shift from small-scale, traditional fermentation to large-scale production processes. In the late 19th and early 20th centuries, industries began to explore the commercial potential of fermentation, especially in food, beverage, and chemical production. During World War I, the demand for acetone, a crucial component in the manufacture of explosives, led to the development of industrial fermentation techniques using bacteria such as *Clostridium acetobutylicum.*

Chaim Weizmann, a chemist and later a prominent political figure, discovered a process to produce acetone and butanol through bacterial fermentation, making a significant contribution to the Allied war effort and showcasing the industrial applications of microbiology. The industrialization of fermentation continued throughout the 20th century, with advancements in bioreactor design, microbial strain selection, and process optimization. As scientists learned to harness specific strains of bacteria and yeast for particular purposes, fermentation became more controlled and predictable. These developments transformed fermentation into a reliable method for producing not only traditional fermented foods but also pharmaceuticals, enzymes, and biofuels. This period marked the transition from fermentation as an art to a science, as industry and academia collaborated to explore the full potential of microbial biotechnology.

1.4 THE RISE OF MODERN BIOTECHNOLOGY

The modern era of fermentation microbiology and biotechnology began in the mid-20th century, with the discovery of DNA's double-helix structure by James Watson and Francis Crick. This discovery laid the groundwork for genetic engineering, which would revolutionize fermentation processes by allowing scientists to modify microbial strains at the genetic level. In the 1970s, the advent of recombinant DNA technology enabled researchers to introduce foreign genes into microorganisms, thus producing recombinant proteins, enzymes, and other valuable biomolecules through fermentation.

One of the most significant applications of modern fermentation biotechnology is the production of insulin. In the 1980s, scientists successfully inserted the gene for human insulin into *Escherichia coli*, a common bacterium, leading to the first biotechnologically produced insulin. This achievement not only improved the availability of insulin for diabetic patients but also demonstrated the power of fermentation technology for pharmaceutical production. Since then, genetically modified organisms (GMOs) have become central to fermentation processes in various industries, including medicine, agriculture, and environmental management.

1.5 FERMENTATION IN THE FOOD AND BEVERAGE INDUSTRY

Despite the industrialization and commercialization of fermentation, its role in food and beverage production remains vital. Traditional fermented foods and beverages, such as yogurt, sauerkraut, kimchi, and kombucha, have experienced a resurgence in popularity due to their health benefits. The probiotic qualities of these foods, which promote gut health and immune function, have made them a staple in health-conscious diets worldwide. Modern fermentation technology has also allowed for the creation of new products, such as plant-based and lab-grown meat alternatives, as consumers seek sustainable and ethical food options.

The rise of craft beer and artisanal food movements in recent decades has further highlighted the cultural and economic importance of fermentation. Small-scale breweries, bakeries, and dairies have popularized traditional fermentation techniques, blending modern microbiology with artisanal practices. This trend has fueled research into novel microbial strains, improved fermentation methods, and the development of high-quality, customized food and beverage products.

1.6 THE FUTURE OF FERMENTATION BIOTECHNOLOGY

The future of fermentation microbiology and biotechnology holds immense promise, particularly in the realms of sustainability, health, and industry. Innovations in synthetic biology, which involves the design and construction of new biological parts and systems, are expected to open new frontiers in fermentation. Scientists are developing genetically engineered microorganisms capable of producing biofuels, biodegradable plastics, and other environmentally friendly materials, presenting a potential solution to the growing problem of pollution and resource depletion.

Moreover, the field of metagenomics, which studies genetic material recovered directly from environmental samples, is enhancing our understanding of microbial communities and their roles in natural fermentation processes. By decoding the genetic information of entire microbial ecosystems, researchers can discover new enzymes, metabolites, and pathways that may be applied to industrial fermentation.

In medicine, fermentation technology is likely to play a significant role in the production of biopharmaceuticals, including vaccines and monoclonal antibodies. As fermentation processes become more sophisticated, the production of complex biological molecules is expected to become faster, more efficient, and more cost-effective. Additionally, advancements in bioprocessing and automation are streamlining the scale-up of fermentation, allowing for rapid and reliable production of high-quality products.

1.7 CONCLUSION

The journey of fermentation microbiology and biotechnology from ancient practices to cutting-edge science exemplifies the remarkable evolution of human understanding and innovation. Once viewed as a mysterious natural phenomenon, fermentation has become a crucial technology with wide-ranging applications in food, medicine, energy, and environmental sustainability. As scientists continue to unravel the complexities of microbial processes and genetic manipulation, fermentation stands poised to remain at the forefront of biotechnology, shaping industries and improving lives in ways that ancient brewers and bakers could scarcely have imagined.

CHAPTER 2

MICROBIOLOGY OF INDUSTRIAL FERMENTATION: CENTRAL AND MODERN CONCEPTS

Industrial fermentation is a complex process involving the controlled use of microorganisms to produce a wide range of products, from pharmaceuticals and biofuels to food and beverages. While fermentation has been practiced for centuries, scientific advancements in microbiology have transformed it into a highly controlled, efficient, and versatile industrial process. The understanding of microbial activity and its optimization for industrial applications has been central to the progress of fermentation technology.

2.1 MICROBIAL BASICS OF INDUSTRIAL FERMENTATION

At its core, industrial fermentation relies on the metabolic capabilities of microorganisms, such as bacteria, yeast, and fungi, to convert organic substances into valuable end products. In fermentation, microorganisms typically break down carbohydrates and other substrates in the absence of oxygen (anaerobic conditions) or with limited oxygen (microaerobic conditions) to produce substances such as ethanol, lactic acid, and other biochemicals. The primary microorganisms used in industrial fermentation are *Saccharomyces cerevisiae* (brewer's yeast), *Lactobacillus* species, *Aspergillus* species, and *Escherichia coli*, among others. Each microorganism has distinct metabolic pathways that yield specific end products, making the selection of strains crucial to the fermentation process.

Different types of fermentation processes can be categorized based on the metabolic pathways and microorganisms involved. For instance, lactic acid fermentation, carried out by lactic acid bacteria, is used in the dairy industry to produce yogurt and cheese. Alcoholic fermentation, driven by yeast, is the basis for producing alcoholic beverages and bioethanol. By understanding the specific metabolic activities of these microorganisms, industries can manipulate fermentation conditions to achieve desired products with maximum efficiency.

2.2 THE ROLE OF BIOREACTORS IN INDUSTRIAL FERMENTATION

Bioreactors are vessels that provide optimal conditions for microbial growth and product formation during industrial fermentation. They are designed to control parameters such as temperature, pH, oxygen concentration, and agitation, ensuring that microorganisms thrive and remain active throughout the process. Modern bioreactors can be highly sophisticated, equipped with sensors and control systems that continuously monitor and adjust conditions in real-time. This automation minimizes the risk of contamination and enhances the reproducibility of fermentation processes, enabling large-scale production.

Bioreactor designs vary according to the type of fermentation process, with batch, fed-batch, and continuous fermentation being the most common. Batch fermentation involves inoculating the bioreactor with microorganisms and substrates at the beginning of the process and allowing fermentation to proceed until completion. In fed-batch fermentation, substrates are added in intervals, extending the fermentation time and increasing yield. Continuous fermentation, where fresh substrate is continuously supplied and products are removed, allows for uninterrupted production, making it suitable for large-scale industrial applications. The choice of bioreactor and fermentation mode depends on factors like product type, production scale, and microorganism characteristics.

2.3 MICROBIAL STRAIN SELECTION AND GENETIC ENGINEERING

One of the central concepts in industrial fermentation is selecting the right microbial strain. Different microorganisms produce different metabolites, and even strains of the same species can vary in productivity, robustness, and tolerance to fermentation conditions. The process of strain selection involves screening various strains to identify those that are best suited to the specific industrial process. In recent years, genetic engineering has revolutionized microbial strain selection, allowing scientists to modify the genetic makeup of microorganisms to enhance desirable traits.

Recombinant DNA technology, for example, enables the introduction of specific genes into microorganisms, leading to the production of recombinant proteins, enzymes, and other valuable compounds. Genetically engineered strains can have increased resistance to harsh fermentation conditions, improved substrate utilization, and higher product yields. The use of *Escherichia coli* for the production of human insulin and *Saccharomyces cerevisiae* for bioethanol production are classic examples of how genetic engineering has enhanced industrial fermentation. Synthetic biology, an extension of genetic engineering, is further expanding the possibilities by creating entirely new metabolic pathways within microorganisms to produce novel products.

2.4 ADVANCES IN PROCESS OPTIMIZATION

Process optimization is key to maximizing the efficiency and yield of industrial fermentation. By optimizing conditions such as substrate concentration, temperature, pH, and oxygen levels, industries can ensure that microorganisms perform at their peak. Mathematical modeling and computer simulations are now frequently used to predict the behavior of microbial cultures under various conditions, allowing for more precise control over fermentation processes.

Additionally, advances in automation and process control technologies have made it possible to implement real-time monitoring and adjustments in fermentation. Sophisticated sensors can measure critical parameters like dissolved oxygen, carbon dioxide levels, and biomass concentration, enabling operators to fine-tune conditions as needed. This not only improves efficiency but also reduces waste and lowers production costs. As a result, the consistency and quality of fermentation products have significantly improved, benefiting industries that rely on uniform, high-quality outputs, such as pharmaceuticals and food production.

2.5 THE IMPACT OF METABOLIC ENGINEERING

Metabolic engineering is an advanced field within industrial microbiology that focuses on optimizing the metabolic pathways within microorganisms to improve product yields. This involves modifying the microorganism's genetic and regulatory networks to direct resources toward desired end products while minimizing by-products. Metabolic engineering has been instrumental in improving the production of biofuels, bioplastics, and other sustainable materials.

For example, researchers have engineered strains of *Saccharomyces cerevisiae* to increase the production of bioethanol by diverting more metabolic resources toward ethanol production. Similarly, metabolic engineering has enabled the production of bioplastics such as polyhydroxyalkanoates (PHAs) by microorganisms like *Cupriavidus necator*. These advances in metabolic engineering not only increase productivity but also help industries transition toward more sustainable practices by providing alternatives to petrochemical-derived products.

2.6 FERMENTATION'S ROLE IN THE BIOECONOMY

The growing emphasis on sustainability has led to the concept of the bioeconomy, where biological resources, including microbial fermentation, are leveraged to create renewable, eco-friendly products. Industrial fermentation plays a central role in this bioeconomy, providing a means to produce biofuels, biodegradable materials, and other sustainable products. Bioethanol, biodiesel, and biogas are renewable energy sources produced through fermentation, offering alternatives to fossil fuels and reducing greenhouse gas emissions.

In the food industry, microbial fermentation is used to create plant-based and lab-grown meat alternatives, which are gaining popularity as sustainable food sources. By using microorganisms to produce essential nutrients and flavors, companies can create alternatives to traditional meat without the environmental footprint associated with animal agriculture. This shift toward microbial fermentation in food production exemplifies the role of fermentation in promoting sustainability in diverse sectors.

2.7 MODERN CHALLENGES AND FUTURE DIRECTIONS

Despite its many advantages, industrial fermentation faces challenges that continue to spur research and innovation. One major challenge is the risk of contamination, which can disrupt fermentation processes and reduce product quality. Although advances in aseptic techniques and bioreactor design have minimized contamination risks, it remains a concern, particularly in large-scale operations. Another challenge is the high cost of substrates, especially in the biofuel industry, where the price of raw materials can impact profitability.

Future advancements in industrial fermentation are likely to focus on improving microbial resilience, enhancing substrate versatility, and optimizing process efficiency. Innovations in synthetic biology and metabolic engineering hold the promise of developing microorganisms that can utilize a wider range of substrates, including waste materials. For example, lignocellulosic biomass, a by-product of agricultural and forestry industries, has potential as a substrate for biofuel production, provided that microorganisms can be engineered to break down its complex structure.

2.8 CONCLUSION

The microbiology of industrial fermentation has evolved into a sophisticated field, with central concepts such as microbial strain selection, bioreactor design, genetic engineering, and process optimization forming the foundation of modern practices. As industries increasingly prioritize sustainability, the role of industrial fermentation in producing renewable and eco-friendly products continues to expand. Advances in metabolic engineering and synthetic biology are likely to further enhance the capabilities of microorganisms, enabling industries to produce a wider variety of products with greater efficiency and lower environmental impact. The future of industrial fermentation lies at the intersection of biology and technology, promising a new era of sustainable, innovative production methods that benefit both industry and society.

CHAPTER 3

FERMENTATION KINETICS: CENTRAL AND MODERN CONCEPTS

Fermentation kinetics is the study of the rates of microbial reactions and the factors influencing them in fermentation processes. Understanding fermentation kinetics is essential for optimizing industrial fermentation systems, as it allows for the control of variables that influence growth rates, product formation, and resource utilization. By mastering these concepts, industries can maximize product yields, reduce production costs, and improve overall process efficiency.

3.1 THE BASICS OF FERMENTATION KINETICS

Fermentation kinetics encompasses several key principles that describe microbial growth, substrate utilization, and product formation. In any fermentation process, microorganisms utilize substrates, typically carbohydrates or other organic compounds, as energy sources. Through metabolic processes, these substrates are transformed into desired products such as ethanol, lactic acid, or other bioactive compounds. Kinetic models aim to describe the relationships among these variables over time, allowing for the prediction of microbial behavior under varying conditions.

The primary parameters in fermentation kinetics are microbial growth rate, substrate consumption rate, and product formation rate. Microbial growth rate, often represented as μ, reflects how quickly a microbial population expands under given conditions. Substrate consumption rate (q_s) measures how rapidly the microorganisms consume available substrates, while product formation rate (q_p) captures the speed at which the desired product is generated. Understanding and optimizing these rates is crucial in industrial fermentation, as it directly impacts yield, efficiency, and cost-effectiveness.

3.2 PHASES OF MICROBIAL GROWTH

Microbial growth in fermentation typically progresses through four distinct phases: lag, exponential (log), stationary, and death. The lag phase represents a period of adaptation when microorganisms adjust to new environmental conditions, preparing for active growth. During the exponential phase, microorganisms grow at their maximum specific growth rate, resulting in rapid substrate consumption and product formation. This phase is particularly important in industrial settings, as it is where the most significant production occurs.

The stationary phase follows, where nutrient depletion or the accumulation of inhibitory by-products slows microbial growth. During this phase, the rate of cell death equals the rate of new cell formation, leading to a stable microbial population.

Finally, the death phase sees a decline in the microbial population as resources are exhausted, and toxic by-products accumulate. In industrial fermentation, managing these growth phases effectively is critical to maintaining high product yields and reducing waste.

3.3 SUBSTRATE-LIMITED KINETICS AND THE MONOD EQUATION

A central concept in fermentation kinetics is substrate-limited growth, where the rate of microbial growth depends on the concentration of the limiting substrate. This relationship is captured by the Monod equation, which is analogous to the Michaelis-Menten equation in enzyme kinetics. The Monod equation describes how microbial growth rate changes in response to substrate concentration:

$$\mu = \frac{\mu_{\max} \cdot S}{K_s + S}$$

In this equation, μ represents the specific growth rate, μ_max is the maximum specific growth rate, S is the substrate concentration, and K_s is the half-saturation constant (the substrate concentration at which the growth rate is half of μ_max). As substrate concentration increases, the growth rate approaches μ_max, but if the substrate concentration is low, growth is limited. This equation is widely used in fermentation kinetics to model microbial growth and optimize substrate utilization.

Understanding substrate-limited kinetics is crucial for managing feed rates in fed-batch and continuous fermentation systems, where substrates are supplied at controlled rates to maximize growth and production without overloading the system or causing inhibition. The Monod equation has proven invaluable for designing and optimizing large-scale industrial fermentation processes, where substrate efficiency directly impacts productivity and cost.

3.4 PRODUCT FORMATION KINETICS: TYPES OF FERMENTATION

In industrial fermentation, product formation kinetics can be categorized into three types: growth-associated, non-growth-associated, and mixed-growth-associated production. Growth-associated production occurs when product formation is directly proportional to microbial growth. Ethanol production by *Saccharomyces cerevisiae* is a classic example, where the primary product is produced during active microbial growth. In such cases, maximizing the exponential growth phase leads to higher product yields.

Non-growth-associated production takes place when products are formed independently of cell growth, often in the stationary phase. For instance, antibiotics produced by *Streptomyces* species are typically synthesized once the microbial population reaches the stationary phase. Mixed-growth-associated production involves product formation during both growth and non-growth phases, as seen in the production of some organic acids and secondary metabolites.

The ability to identify the type of product formation kinetics is essential for optimizing fermentation processes. By adjusting the timing and rate of substrate addition, as well as controlling environmental factors like pH and temperature, operators can drive the fermentation towards higher productivity based on the specific kinetic profile of the desired product.

3.5 ADVANCES IN MODELING FERMENTATION KINETICS

Recent advancements in computational biology and data science have led to the development of more sophisticated models for fermentation kinetics. Traditional models, such as the Monod equation, are useful for simple fermentation processes but often lack the complexity to account for dynamic, multi-phase interactions in large-scale operations. Modern kinetic models incorporate factors like environmental changes, cellular metabolism, and the effects of genetic modifications on microbial behavior.

Dynamic models, for instance, use differential equations to capture the time-dependent behavior of microbial populations, substrate concentrations, and product formation rates. These models can simulate fermentation processes under various conditions, allowing for predictive analysis and process optimization. Machine learning algorithms are also increasingly used to model complex fermentation kinetics, with the ability to analyze large datasets and recognize patterns that may not be apparent through traditional approaches. This technology enables the creation of more accurate, real-time models that can adapt to fluctuations in fermentation conditions.

3.6 KINETIC CONTROL IN FED-BATCH AND CONTINUOUS FERMENTATION

Fed-batch and continuous fermentation are two common methods used to control kinetic parameters in industrial settings. In fed-batch fermentation, substrates are periodically added to the bioreactor, extending the exponential growth phase and increasing product yield. By controlling substrate feed rates based on kinetic models, industries can maintain optimal growth and production conditions for extended periods. This method is especially useful in cases where the product formation is growth-associated, as it prevents the depletion of substrates and prolongs the productive phase.

Continuous fermentation involves the continuous addition of substrates and the removal of products, maintaining a steady-state condition. In continuous processes, the rate of substrate addition and product removal can be adjusted to keep the microbial population in the exponential phase, maximizing productivity. The challenge with continuous fermentation is maintaining microbial stability, as contamination or genetic drift can impact productivity. Kinetic modeling and real-time monitoring are therefore essential for maintaining stable, long-term continuous fermentation operations.

3.7 THE ROLE OF KINETICS IN METABOLIC ENGINEERING AND SYNTHETIC BIOLOGY

Metabolic engineering and synthetic biology have introduced new dimensions to fermentation kinetics by allowing for precise modifications of microbial metabolic pathways. These modifications can alter the kinetics of substrate utilization and product formation, enabling the production of non-native products or enhancing yields of specific metabolites. Through genetic engineering, scientists can introduce or knock out genes to optimize metabolic fluxes toward desired end products.

For instance, modifying *Escherichia coli* to produce high levels of succinic acid—a valuable bio-based chemical—requires the re-routing of metabolic pathways and the introduction of new enzymes. By altering the kinetics of these engineered pathways, scientists can increase the efficiency of substrate-to-product conversion. The integration of kinetic models with genetic engineering and synthetic biology approaches is becoming a powerful tool for designing microorganisms that can achieve industrial production targets with high efficiency and minimal waste.

3.8 FUTURE DIRECTIONS IN FERMENTATION KINETICS

The field of fermentation kinetics is poised for significant advancements as industries adopt new technologies and approaches. One promising area is the use of real-time monitoring and control systems that employ artificial intelligence (AI) to optimize fermentation processes on the fly. These systems use data from sensors and kinetic models to make instant adjustments to fermentation conditions, ensuring maximum productivity and minimal waste.

Another area of development is the use of CRISPR-based gene editing to create microorganisms with customized kinetic profiles. By precisely editing genes associated with growth, substrate utilization, and product formation, scientists can create strains tailored to specific fermentation processes. Additionally, advances in single-cell analysis are allowing researchers to study the kinetics of individual cells within a population, providing insights into microbial heterogeneity and enabling the design of more consistent fermentation processes.

3.9 CONCLUSION

Fermentation kinetics provides a foundation for understanding and optimizing the complex interactions between microorganisms, substrates, and environmental conditions in industrial fermentation. Traditional kinetic models, such as the Monod equation, have proven invaluable in guiding process design and optimization.

However, modern advancements in modeling, real-time monitoring, and genetic engineering are transforming the field, enabling industries to achieve unprecedented levels of efficiency and control. As fermentation kinetics continues to evolve, it will play an increasingly central role in biotechnological applications, driving innovation in sectors ranging from food production to pharmaceuticals and sustainable biofuels.

SECTION II:

FERMENTATION MICROBIOLOGY AND BIOTECHNOLOGY: CONTROL OF CARBON FLUX TO PRODUCT FORMATION IN MICROBIAL CELL FACTORIES

CHAPTER 4

MICROBIAL SYNTHESIS OF PRIMARY METABOLITES: CURRENT TRENDS AND FUTURE PROSPECTS

The microbial synthesis of primary metabolites has garnered significant interest due to its potential to produce essential compounds sustainably and efficiently. Primary metabolites are substances that are crucial for microbial growth and metabolism, including amino acids, nucleotides, vitamins, and organic acids.

Their applications span numerous industries, from food and pharmaceuticals to agriculture and biofuels. Recent advancements in biotechnology have enhanced the production of primary metabolites, leveraging microbial pathways, genetic engineering, and synthetic biology.

4.1 UNDERSTANDING PRIMARY METABOLITES AND THEIR INDUSTRIAL SIGNIFICANCE

Primary metabolites are compounds that microbes synthesize during their growth phase as essential components for their cellular processes. Unlike secondary metabolites, which are often species-specific and accumulate under nutrient-limited conditions, primary metabolites are universally produced across species during exponential growth.

They serve as building blocks for cellular structures and functions, including amino acids for protein synthesis, nucleotides for DNA and RNA, and vitamins for coenzymatic functions.

Industrially, primary metabolites are valuable for several reasons. Amino acids, for instance, are used extensively in the food and pharmaceutical sectors. Organic acids, such as lactic acid and citric acid, are widely applied as preservatives and acidulants.

With the rising demand for sustainable, eco-friendly production processes, microbial synthesis offers an attractive alternative to traditional chemical synthesis, as it often requires milder conditions and produces fewer by-products. Additionally, microbes can often utilize renewable feedstocks, making microbial synthesis a viable option for industries aiming to reduce environmental impact.

4.2 THE ROLE OF GENETIC ENGINEERING IN METABOLITE PRODUCTION

Genetic engineering has transformed microbial synthesis by allowing for precise manipulation of microbial metabolic pathways. Using techniques such as recombinant DNA technology and CRISPR-Cas9, scientists can modify the genetic structure of microorganisms to enhance the production of primary metabolites. By introducing or overexpressing specific genes, researchers can amplify pathways that lead to increased yields of desired metabolites.

For example, the production of amino acids such as lysine and glutamate has been optimized through genetic engineering. *Corynebacterium glutamicum*, a microbe known for its ability to produce amino acids, has been modified to overproduce lysine by targeting metabolic regulators and blocking competing pathways. Additionally, researchers have utilized gene editing to delete genes associated with feedback inhibition, a process that often limits metabolite accumulation. These genetic modifications have resulted in strains capable of producing amino acids in much higher quantities than wild-type strains.

CRISPR technology has further expanded possibilities by enabling the rapid and precise editing of multiple genes simultaneously. In the case of nucleotide production, researchers have applied CRISPR to modify genes in *Escherichia coli*, boosting the synthesis of nucleotides such as inosine and guanosine, which are used as flavor enhancers in the food industry. By fine-tuning regulatory networks, these engineered strains achieve higher yields, reducing the production cost and improving feasibility for commercial applications.

4.3 ADVANCES IN SYNTHETIC BIOLOGY AND PATHWAY ENGINEERING

Synthetic biology has opened new frontiers in microbial synthesis, particularly in designing novel metabolic pathways that do not naturally exist in the target microorganisms.

Pathway engineering is a process in synthetic biology where researchers design and introduce artificial metabolic pathways into microbes to produce specific primary metabolites. This approach has made it possible to utilize diverse substrates, including waste products and alternative carbon sources, expanding the scope of microbial synthesis.

One notable example is the microbial synthesis of succinic acid, an organic acid used in bioplastics, solvents, and other industrial applications. Researchers have successfully introduced synthetic pathways into *Escherichia coli* and *Saccharomyces cerevisiae* to increase succinic acid production by redirecting metabolic flux towards the tricarboxylic acid (TCA) cycle. By introducing foreign genes and optimizing regulatory elements, these engineered strains can convert sugars and glycerol into succinic acid more efficiently. Another promising area in synthetic biology is the development of modular biosynthetic pathways. Researchers create and assemble standardized genetic modules that can be easily inserted or removed, allowing for the flexible production of various metabolites. Modular pathways are particularly useful in optimizing the production of vitamins and cofactors, where balancing intermediate steps is essential. Through synthetic biology, researchers can fine-tune each step to ensure that metabolite production is optimized at every stage.

4.4 METABOLIC ENGINEERING AND FLUX ANALYSIS FOR OPTIMIZED PRODUCTION

Metabolic engineering involves the modification of cellular networks to redirect metabolic flux towards desired end products. Flux analysis is a powerful tool in this approach, as it enables researchers to track the flow of carbon, nitrogen, and other key elements through cellular pathways. By understanding these flows, scientists can identify bottlenecks, assess energy requirements, and rewire cellular pathways for maximum efficiency.

Using flux balance analysis (FBA), researchers have optimized the microbial production of lactic acid, which is widely used in the food, cosmetic, and bioplastic industries. Through metabolic engineering, scientists increased the carbon flux through glycolysis while reducing by-products, leading to higher lactic acid yields in bacteria like *Lactobacillus* and *Escherichia coli*. Advances in computational biology have further enhanced the accuracy of FBA, making it a reliable method for pathway optimization.

The production of other high-value compounds, such as coenzyme Q10 (an essential molecule in cellular respiration), has also benefited from metabolic engineering. By using flux analysis to assess coenzyme biosynthesis pathways in yeast and bacteria, researchers have identified key enzymes that can be upregulated to increase production. These modifications not only boost yield but also reduce the overall costs associated with extraction and purification.

4.5 TRENDS IN FERMENTATION TECHNOLOGY AND BIOREACTOR DESIGN

Fermentation technology has evolved significantly, with modern bioreactors enabling precise control over environmental factors such as pH, temperature, oxygen levels, and substrate concentration. In primary metabolite production, achieving optimal fermentation conditions is crucial, as small fluctuations can impact product yield and quality. Recent trends in bioreactor design focus on real-time monitoring and adaptive control systems that automatically adjust parameters based on microbial behavior.

Continuous fermentation and fed-batch fermentation are two commonly used techniques for metabolite production. Continuous fermentation allows for constant substrate feeding and product removal, maintaining a steady state and maximizing productivity. In fed-batch systems, substrate is added in intervals to prolong the growth phase, which is beneficial for growth-associated metabolites like amino acids. By integrating advanced sensors and automated control systems, industries can maintain ideal conditions for extended periods, reducing the risk of contamination and enhancing consistency.

Microbial cell factories are another trend in fermentation technology, where researchers engineer microbial strains to act as "factories" that produce specific primary metabolites. These cell factories are designed to convert low-cost substrates into high-value products, offering a sustainable alternative to chemical synthesis. Microbial cell factories are already being used for the production of vitamins, amino acids, and organic acids, showcasing their potential in reducing dependency on traditional manufacturing methods.

4.6 THE FUTURE OF MICROBIAL SYNTHESIS: CHALLENGES AND PROSPECTS

Despite remarkable progress, microbial synthesis of primary metabolites faces several challenges. The complexity of microbial metabolic networks can make it difficult to achieve optimal yields, as modifying one pathway often affects others. Moreover, contamination, genetic instability, and the high costs of substrates are additional obstacles in large-scale production. Efforts are ongoing to address these issues, with advances in gene-editing tools, synthetic biology, and computational modeling expected to provide solutions.

Looking forward, microbial synthesis is poised to play a central role in the bioeconomy, offering eco-friendly production methods for a variety of industries. As researchers develop more robust and versatile microbial strains, the range of possible metabolites will expand, providing alternatives to petroleum-based products. For instance, the production of bio-based polymers, flavor compounds, and nutritional supplements is likely to see continued growth, with microbial synthesis serving as a sustainable and cost-effective pathway.

The integration of artificial intelligence (AI) and machine learning (ML) into fermentation processes is another promising development. By analyzing large datasets from microbial fermentations, AI algorithms can identify optimal conditions, predict outcomes, and even suggest genetic modifications. Such predictive capabilities will allow for greater efficiency, reduced waste, and improved scalability, making microbial synthesis more competitive with traditional manufacturing.

4.7 CONCLUSION

Microbial synthesis of primary metabolites has emerged as a cornerstone of industrial biotechnology, with applications ranging from food and pharmaceuticals to sustainable materials. Advances in genetic engineering, synthetic biology, metabolic engineering, and fermentation technology have collectively driven significant progress, enabling more efficient and sustainable production of essential compounds.

While challenges remain, particularly in scaling up and optimizing microbial strains, the future of microbial synthesis is bright. With ongoing research and technological innovation, microbial synthesis has the potential to transform industries by providing eco-friendly, scalable alternatives to conventional production methods, contributing to a more sustainable future.

CHAPTER 5

MICROBIAL SYNTHESIS OF SECONDARY METABOLITES AND STRAIN IMPROVEMENT: CURRENT TRENDS AND FUTURE PROSPECTS

Secondary metabolites, often referred to as "natural products," are organic compounds produced by microorganisms that are not essential for their growth or survival but play a significant role in the organism's adaptability, interaction, and competition within ecosystems. Unlike primary metabolites, secondary metabolites are typically produced during the stationary phase of growth and can have complex chemical structures. These compounds have enormous industrial value as antibiotics, anticancer agents, immunosuppressants, and enzyme inhibitors. Due to their bioactive properties, the microbial synthesis of secondary metabolites and the improvement of microbial strains to enhance their production are critical in pharmaceuticals, agriculture, and biotechnology.

5.1 UNDERSTANDING THE SECONDARY METABOLITES

Secondary metabolites, also called specialised metabolites or secondary products, are a type of natural product (organic compounds produced by living organisms) generated by lifeforms (e.g., bacteria, archaea, fungi, plants, or animals) that are not directly involved in the normal growth, development, or reproduction of the organism. Instead, they generally mediate ecological interactions, which may produce a selective advantage for the organism by increasing its survivability or fecundity. Specific secondary metabolites are often restricted to a narrow set of species within a phylogenetic group. Secondary metabolites often play an important role in plant defense against herbivory and other interspecies defenses. Humans use secondary metabolites as medicines, flavourings, pigments, and recreational drugs.

The term secondary metabolite was first coined by Albrecht Kossel, the 1910 Nobel Prize laureate for medicine and physiology. 30 years later a Polish botanist Friedrich Czapek described secondary metabolites as end products of nitrogen metabolism.

Secondary metabolites commonly mediate antagonistic interactions, such as competition and predation, as well as mutualistic ones such as pollination and resource mutualisms. Usually, secondary metabolites are confined to a specific lineage or even species, though there is considerable evidence that horizontal transfer across species or genera of entire pathways plays an important role in bacterial (and, likely, fungal) evolution.

Research also shows that secondary metabolism can affect different species in varying ways. In the same forest, four separate species of arboreal marsupial folivores reacted differently to a secondary metabolite in eucalypts. This shows that differing types of secondary metabolites can be the split between two herbivore ecological niches.

Additionally, certain species evolve to resist secondary metabolites and even use them for their own benefit. For example, monarch butterflies have evolved to be able to eat milkweed (*Asclepias*) despite the presence of toxic cardiac glycosides. The butterflies are not only resistant to the toxins, but are actually able to benefit by actively sequestering them, which can lead to the deterrence of predators.

5.1.1 Plant Secondary Metabolites

Plants are capable of producing and synthesizing diverse groups of organic compounds and are divided into two major groups: primary and secondary metabolites. Secondary metabolites are metabolic intermediates or products which are not essential to growth and life of the producing plants but rather required for interaction of plants with their environment and produced in response to stress. Their antibiotic, antifungal and antiviral properties protect the plant from pathogens. Some secondary metabolites such as phenylpropanoids protect plants from UV damage.

The biological effects of plant secondary metabolites on humans have been known since ancient times. The herb *Artemisia annua* which contains Artemisinin, has been widely used in Chinese traditional medicine more than two thousand years ago. Plant secondary metabolites are classified by their chemical structure and can be divided into four major classes: terpenes, phenylpropanoids (i.e. phenolics), polyketides, and alkaloids.

Chemical Classes

Terpenoids

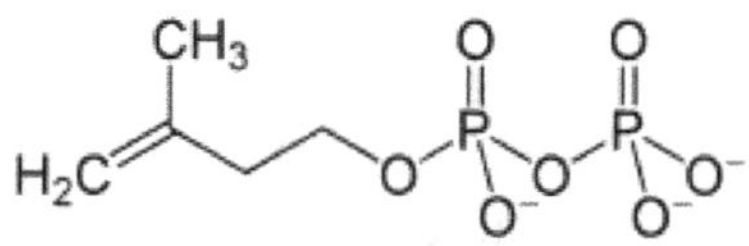

Fig 1: Structural Formula for Isopentenyl Pyrophosphate (IPP)

The carbon atoms of which constitute building blocks in terpenes. Because of the presence of five carbon atoms, the derived building block it is termed a C5 unit.

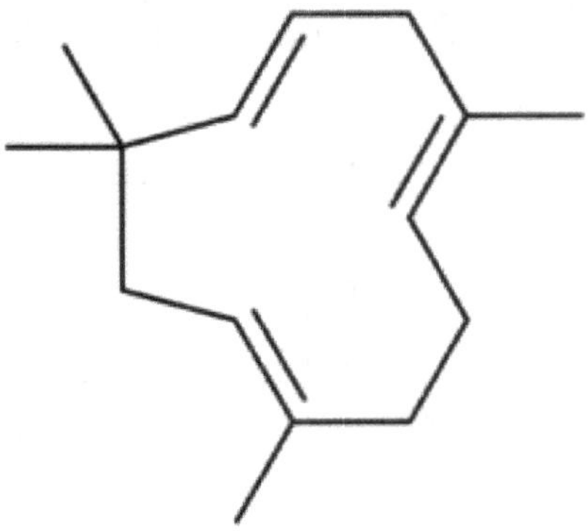

Fig 2: Structural Formula for Humulene, A Monocyclic Sesquiterpene (called so because it Contains 15 Carbon Atoms) that is Built from Three C5 Units Derived from Isopentenyl Pyrophosphate (IPP)

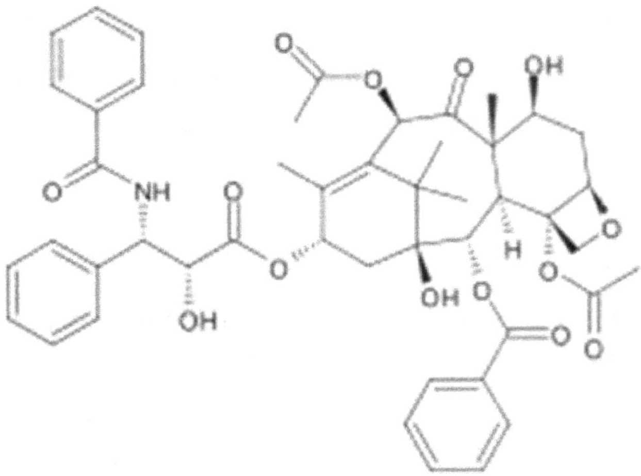

Fig 3: Skeletal Formula of the Terpenoid Taxol, An Anticancer Drug

Terpenes constitute a large class of natural products which are composed of isoprene units. Terpenes are only hydrocarbons and terpenoids are oxygenated hydrocarbons.

The general molecular formula of terpenes are multiples of $(C_5H_8)_{n,}$ where 'n' is number of linked isoprene units. Hence, terpenes are also termed as isoprenoid compounds.

Classification is based on the number of isoprene units present in their structure. Some terpenoids (i.e. many sterols) are primary metabolites.

Some terpenoids that may have originated as secondary metabolites have subsequently been recruited as plant hormones, such as gibberellins, brassinosteroids, and strigolactones.

Number of Isoprene Units	Name	Carbon Atoms
1	Hemiterpene	C_5
2	Monoterpene	C_{10}
3	Sesquiterpenes	C_{15}
4	Diterpene	C_{20}
5	Sesterterpene	C_{25}
6	Triterpene	C_{30}
7	Sesquarterterpene	C_{35}
8	Tetraterpene	C_{40}
More than 8	Polyterpene	

Examples of terpenoids built from hemiterpene oligomerization are:

- Azadirachtin, present in *Azadirachta indica*, the (Neem tree).
- Artemisinin, present in *Artemisia annua*, Chinese wormwood.
- Tetrahydrocannabinol, present in *Cannabis sativa*, cannabis.
- Saponins, glycosylated triterpenes present in e.g. *Chenopodium quinoa*, quinoa.

Phenolic Compounds

Phenolics are a chemical compound characterized by the presence of aromatic ring structure bearing one or more hydroxyl groups.

Phenolics are the most abundant secondary metabolites of plants ranging from simple molecules such as phenolic acid to highly polymerized substances such as tannins. Classes of phenolics have been characterized on the basis of their basic skeleton.

No. of Carbon Atoms	Basic Skeleton	Class
6	C_6	Simple phenols
7	$C_6 - C_1$	Phenolic acids
8	$C_6 - C_2$	Acetophenone, Phenyle acetic acid
9	$C_6 - C_3$	Phenylepropanoids, hydroxycinnamic acid, coumarins
10	$C_6 - C_4$	Naphthoquinone
13	$C_6 - C_1 - C_6$	Xanthone
14	$C_6 - C_2 - C_6$	Stilbene, anthraquinone
15	$C_6 - C_3 - C_6$	Flavonoids, isoflavanoids
18	$(C_6 - C_3)_2$	lignans, neolignans
30	$(C_6 - C_3 - C_6)_2$	Biflavonoids

An example of a plant phenol is:

- Resveratrol, a C_{14} stilbenoid produced by e.g. grapes.

Alkaloids

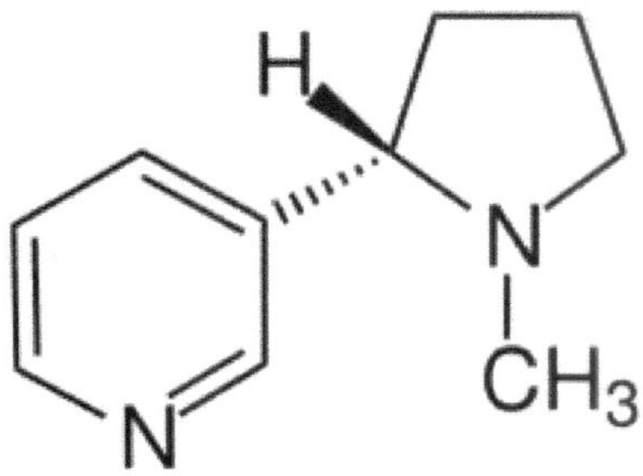

Fig 4: Structural Formula for the Alkaloid Nicotine

The structure includes two nitrogen atoms both of which derives from amino acids. The nitrogen atom in the pyridine ring derives from aspartate, whereas that in the pyrrolidine ring derives from arginine (or ornithine).

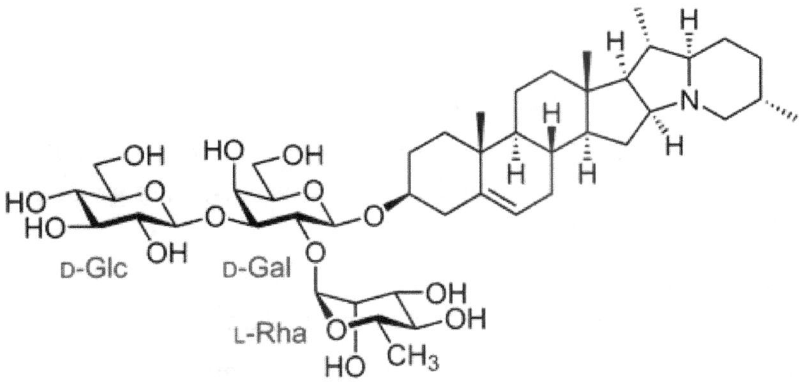

Fig 5: Skeletal Formula of Solanine, A Toxic Alkaloid which Builds up in Potatoes

Alkaloids are a diverse group of nitrogen-containing basic compounds. They are typically derived from plant sources and contain one or more nitrogen atoms. Chemically they are very heterogeneous. Based on chemical structures, they may be classified into two broad categories:

Non heterocyclic or atypical alkaloids, for example hordenine or *N*-methyltyramine, colchicine, and taxol

Heterocyclic or typical alkaloids, for example quinine, caffeine, and nicotine

Examples of alkaloids produced by plants are:

- Hyoscyamine, present in *Datura stramonium.*
- Atropine, present in *Atropa belladonna*, deadly nightshade.
- Cocaine, present in *Erythroxylum coca* the Coca plant.

- Scopolamine, present in the *Solanaceae* (nightshade) plant family.
- Codeine and morphine, present in *Papaver somniferum*, the opium poppy.
- Vincristine and vinblastine, mitotic inhibitors found in *Catharanthus roseus*, the rosy periwinkle.

Many alkaloids affect the central nervous system of animals by binding to neurotransmitter receptors.

Glucosinolates

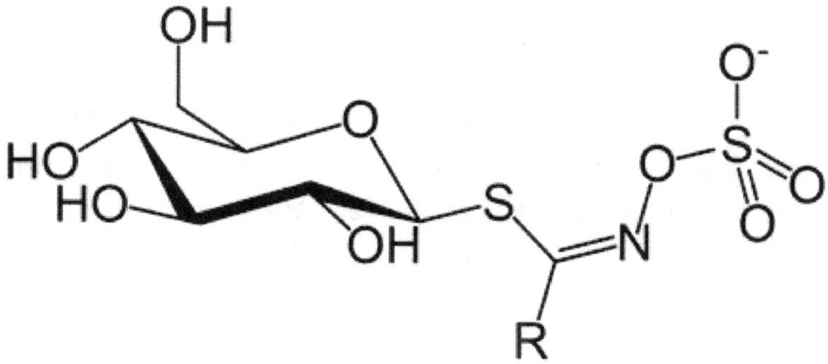

Fig 6: Structural Formula for Glucosinolates

The side group *R* can vary. The structure includes a glucose molecules, a nitrogen atom derived from an amino acid, and two sulfur atoms, among which one derives from glutathione and the other from sulfate.

Glucosinolates are secondary metabolites that include both sulfur and nitrogen atoms, and are derived from glucose, an amino acid and sulfate.

An example of a glucosinolate in plants is Glucoraphanin, from broccoli (*Brassica oleracea* var. *italica*).

Plant Secondary Metabolites in Medicine

Many drugs used in modern medicine are derived from plant secondary metabolites. The two most commonly known terpenoids are artemisinin and paclitaxel. Artemisinin was widely used in Traditional Chinese medicine and later rediscovered as a powerful antimalarial by a Chinese scientist Tu Youyou. She was later awarded the Nobel Prize for the discovery in 2015. Currently, the malaria parasite, *Plasmodium falciparum*, has become resistant to artemisinin alone and the World Health Organization recommends its use with other antimalarial drugs for a successful therapy. Paclitaxel the active compound found in Taxol is a chemotherapy drug used to treat many forms of cancers including ovarian cancer, breast cancer, lung cancer, Kaposi sarcoma, cervical cancer, and pancreatic cancer. Taxol was first isolated in 1973 from barks of a coniferous tree, the Pacific Yew.

Morphine and codeine both belong to the class of alkaloids and are derived from opium poppies. Morphine was discovered in 1804 by a German pharmacist Friedrich Sertürnert.

It was the first active alkaloid extracted from the opium poppy. It is mostly known for its strong analgesic effects, however, morphine is also used to treat shortness of breath and treatment of addiction to stronger opiates such as heroin. Despite its positive effects on humans, morphine has very strong adverse effects, such as addiction, hormone imbalance or constipation. Due to its highly addictive nature morphine is a strictly controlled substance around the world, used only in very severe cases with some countries underusing it compared to the global average due to the social stigma around it.

Codeine, also an alkaloid derived from the opium poppy, is considered the most widely used drug in the world according to World Health Organization. It was first isolated in 1832 by a French chemist Pierre Jean Robiquet, also known for the discovery of caffeine and a widely used red dye alizarin. Primarily codeine is used to treat mild pain and relief coughing although in some cases it is used to treat diarrhea and some forms of irritable bowel syndrome. Codeine has the strength of 0.1-0.15 compared to morphine ingested orally, hence it is much safer to use. Although codeine can be extracted from the opium poppy, the process is not feasible economically due to the low abundance of pure codeine in the plant. A chemical process of methylation of the much more abundant morphine is the main method of production.

Atropine is an alkaloid first found in *Atropa belladonna*, a member of the nightshade family. While atropine was first isolated in the 19th century, its medical use dates back to at least the fourth century B.C., where it was used for wounds, gout, and sleeplessness. Currently atropine is administered intravenously to treat bradycardia and as an antidote to organophosphate poisoning. Overdosing of atropine may lead to atropine poisoning which results in side effects such as blurred vision, nausea, lack of sweating, dry mouth and tachycardia.

Resveratrol is a phenolic compound of the flavonoid class. It is highly abundant in grapes, blueberries, raspberries and peanuts. It is commonly taken as a dietary supplement for extending life and reducing the risk of cancer and heart disease, however there is no strong evidence supporting its efficacy. Nevertheless, flavonoids are in general thought to have beneficial effects for humans. Certain studies shown that flavonoids have direct antibiotic activity. A number of in vitro and limited in vivo studies shown that flavonoids such as quercetin have synergistic activity with antibiotics and are able to suppress bacterial loads.

Digoxin is a cardiac glycoside first derived by William Withering in 1785 from the foxglove *(Digitalis)* plant. It is typically used to treat heart conditions such as atrial fibrillation, atrial flutter or heart failure. Digoxin can, however, have side effects such as nausea, bradycardia, diarrhea or even life-threatening arrhythmia.

5.1.2 Fungal Secondary Metabolites

The three main classes of fungal secondary metabolites are: polyketides, nonribosomal peptides and terpenes. Although fungal secondary metabolites are not required for growth they play an essential role in survival of fungi in their ecological niche. The most known fungal secondary metabolite is penicillin discovered by Alexander Fleming in 1928. Later in 1945, Fleming, alongside Ernst Chain and Howard Florey, received a Nobel Prize for its discovery which was pivotal in reducing the number of deaths in World War II by over 100,000.

Examples of other fungal secondary metabolites are:

- Lovastatin, a polyketide from e.g. *Pleurotus ostreatus*, oyster mushrooms.
- Aflatoxin B1, a polyketide from *Aspergillus flavus.*
- Ciclosporin, a non-ribosomal cyclic peptide from *Tolypocladium inflatum.*

Lovastatin was the first FDA approved secondary metabolite to lower cholesterol levels. Lovastatin occurs naturally in low concentrations in oyster mushrooms, red yeast rice, and Pu-erh. Lovastatin's mode of action is competitive inhibition of HMG-CoA reductase, and a rate-limiting enzyme responsible for converting HMG-CoA to mevalonate.

Fungal secondary metabolites can also be dangerous to humans. *Claviceps purpurea,* a member of the ergot group of fungi typically growing on rye, results in death when ingested.

The build-up of poisonous alkaloids found in *C. purpurea* lead to symptoms such as seizures and spasms, diarrhea, paresthesias, Itching, psychosis or gangrene. Currently, removal of ergot bodies requires putting the rye in brine solution with healthy grains sinking and infected floating.

5.1.3 Bacterial Secondary Metabolites

Bacterial production of secondary metabolites starts in the stationary phase as a consequence of lack of nutrients or in response to environmental stress. Secondary metabolite synthesis in bacteria is not essential for their growth, however, they allow them to better interact with their ecological niche.

The main synthetic pathways of secondary metabolite production in bacteria are; b-lactam, oligosaccharide, shikimate, polyketide and non-ribosomal pathways. Many bacterial secondary metabolites are toxic to mammals. When secreted those poisonous compounds are known as exotoxins whereas those found in the prokaryotic cell wall are endotoxins.

Examples of bacterial secondary metabolites are:

Phenazine

- Pyocyanin, from *Pseudomonas aeruginosa.*
- Other phenazines from *Pseudomonas* ssp. and *Streptomyces* ssp.

Polyketides

- Avermectin, from *Streptomyces avermitilis.*
- Epothilones, macrolactones from the soil-dwelling myxobacterium *Sorangium cellulosum.*
- Erythromycin, *Saccharopolyspora erythraea.*
- Nystatin, from *Streptomyces noursei.*
- Rifamycin, from *Amycolatopsis rifamycinica.*

Nonribosomal Peptides

- Bacitracin, from *Bacillus subtilis* (Tracy strain).
- Gramicidin, from *Brevibacillus brevis.*
- Polymyxin, from *Paenibacillus polymyxa.*
- Ramoplanin, from *Actinoplanes* strain ATCC 33076.
- Teicoplanins, from *Actinoplanes teicomyceticus.*
- Vancomycin, from the soil bacterium *Amycolatopsis orientalis.*

Ribosomal Peptides

- Microcins, bacteriocins such as microcin V from *Escherichia coli.*
- Thiostrepton, from several strains of streptomycetes, e.g. *Streptomyces azureus.*

Glucosides

- Nojirimycin, an iminosugar from a class of *Streptomyces* species.

Alkaloids

- Tetrodotoxin, a neurotoxin produced by *Pseudoalteromonas* and other bacteria living in symbiosis with animals such as e.g. pufferfish.

Terpenoids

- Carotenoids, a pigment produced by different species of bacteria, such as *Micrococcus* sp.

5.1.4 Archaea Secondary Metabolites

Archaea are capable of producing a variety of secondary metabolites, which may have significant biotechnological applications. Despite knowing this, the biosynthetic pathways for secondary metabolites in archaea are less understood than those in bacteria. Notably, archaea often lack some biosynthesis genes commonly present in bacteria, which suggests that they may possess unique metabolic pathways for synthetizing these compounds.

Extracellular Polymeric Substances

Extracellular polymeric substances can effectively adsorb and degrade hazardous organic chemicals. While these compounds are produced by various organisms, archaea are particularly promising for wastewater treatment due to their high tolerance to saline concentrations and their ability to grow anaerobically.

5.1.5 Biotechnological Approaches

Selective breeding was used as one of the first biotechnological techniques used to reduce the unwanted secondary metabolites in food, such as naringin causing bitterness in grapefruit. In some cases increasing the content of secondary metabolites in a plant is the desired outcome. Traditionally this was done using in-vitro plant tissue culture techniques which allow for: control of growth conditions, mitigate seasonality of plants or protect them from parasites and harmful-microbes. Synthesis of secondary metabolites can be further enhanced by introducing elicitors into a tissue plant culture, such as jasmonic acid, UV-B or ozone. These compounds induce stress onto a plant leading to increased production of secondary metabolites.

To further increase the yield of SMs new approaches have been developed. A novel approach used by Evolva uses recombinant yeast S. cerevisiae strains to produce secondary metabolites normally found in plants.

The first successful chemical compound synthesised with Evolva was vanillin, widely used in the food beverage industry as flavouring. The process involves inserting the desired secondary metabolite gene into an artificial chromosome in the recombinant yeast leading to synthesis of vanillin. Currently Evolva produces a wide array of chemicals such as stevia, resveratrol or nootkatone.

Nagoya Protocol

With the development of recombinant technologies the *Nagoya Protocol on Access to Genetic Resources and the Fair and Equitable Sharing of Benefits Arising from their Utilization to the Convention on Biological Diversity* was signed in 2010. The protocol regulates the conservation and protection of genetic resources to prevent the exploitation of smaller and poorer countries. If genetic, protein or small molecule resources sourced from biodiverse countries become profitable a compensation scheme was put in place for the countries of origin.

5.2 THE ROLE OF SECONDARY METABOLITES IN INDUSTRY

Secondary metabolites produced by microbes include antibiotics, pigments, toxins, alkaloids, and other complex molecules. Antibiotics, for example, are derived from secondary metabolites synthesized by fungi and bacteria, primarily *Streptomyces* and *Penicillium* species. Compounds like penicillin, streptomycin, and tetracycline have become the cornerstone of modern medicine, revolutionizing the treatment of bacterial infections. Other secondary metabolites, such as cyclosporine, used as an immunosuppressant, and lovastatin, a cholesterol-lowering agent, also have wide applications.

In agriculture, secondary metabolites like insecticides, fungicides, and herbicides derived from microbes are used as natural biopesticides. The production of these metabolites by microorganisms provides an eco-friendly alternative to synthetic chemicals, which are often harmful to the environment and can cause resistance in pests and pathogens. Given their diverse applications, the demand for microbial-derived secondary metabolites continues to grow, driving research on methods to enhance and optimize their production.

5.3 CURRENT TRENDS IN MICROBIAL SYNTHESIS OF SECONDARY METABOLITES

The production of secondary metabolites in microbes is often regulated by complex networks of genes and is affected by environmental conditions. Modern research in microbial synthesis focuses on optimizing production through genetic and metabolic engineering techniques to improve yield and efficiency. Strategies such as the use of precursor feeding, optimization of culture conditions, and co-culture fermentation have shown promising results in increasing the yield of secondary metabolites.

One trend gaining momentum is the use of co-culturing, which involves growing two or more microorganisms together to stimulate the production of secondary metabolites. In natural environments, microbes often exist in communities and interact with each other, leading to the activation of otherwise silent or lowly expressed biosynthetic pathways. Co-culturing mimics these interactions, leading to the production of unique compounds that may not be synthesized in monocultures. For instance, co-culturing *Streptomyces* species with *Bacillus* has been shown to enhance antibiotic production by triggering otherwise inactive gene clusters.

The use of elicitors, substances that stimulate secondary metabolite production, is another emerging strategy. Elicitors can be physical, chemical, or biological agents that mimic environmental stress, triggering pathways associated with secondary metabolite synthesis. These may include UV light, specific nutrient limitations, or microbial signals. For instance, the application of UV irradiation on certain fungal cultures can enhance the production of pigments and bioactive compounds, allowing industries to obtain desired compounds more efficiently.

5.4 STRAIN IMPROVEMENT TECHNIQUES: GENETIC ENGINEERING AND METABOLIC PATHWAY OPTIMIZATION

The ability to improve microbial strains is crucial for enhancing secondary metabolite production, as it allows for the modification of metabolic pathways and the removal of bottlenecks in production. Genetic engineering, which involves the targeted modification of genes within microorganisms, has enabled significant advancements in strain improvement. Techniques like CRISPR-Cas9 gene editing, RNA interference, and recombineering allow scientists to modify genes associated with regulatory mechanisms in secondary metabolism.

CRISPR-Cas9 has emerged as a revolutionary tool in strain improvement for its precision in gene editing. By targeting specific gene sequences, scientists can activate or repress pathways responsible for secondary metabolite synthesis. For example, in *Aspergillus niger*, CRISPR-Cas9 has been used to delete genes encoding for undesired by-products, thereby increasing the production of desired secondary metabolites. Similarly, CRISPR has been used in *Streptomyces* species to remove genes that control feedback inhibition, enabling a continuous production of antibiotics.

Another effective technique is pathway engineering, where entire biosynthetic pathways are modified to optimize metabolite production. By identifying rate-limiting steps and redirecting metabolic flux toward the desired compound, pathway engineering can enhance the productivity of a target secondary metabolite. In *Penicillium chrysogenum*, for example, pathway engineering has led to optimized production of penicillin by increasing the flow of precursor molecules and reducing unwanted side reactions.

5.5 SYNTHETIC BIOLOGY AND THE DISCOVERY OF NOVEL SECONDARY METABOLITES

Synthetic biology goes beyond traditional genetic engineering by allowing scientists to construct new biosynthetic pathways or even create synthetic microbes capable of producing high yields of specific secondary metabolites. With synthetic biology, scientists can introduce entire gene clusters from one organism into another, enabling non-native synthesis. For instance, the gene cluster responsible for producing erythromycin, an antibiotic originally derived from *Saccharopolyspora erythraea*, has been transferred into *Escherichia coli* for enhanced production.

Another advancement in synthetic biology is the development of cell-free systems for secondary metabolite synthesis. In cell-free systems, enzymes from microbes are isolated and used in reactions outside of living cells. This system allows for direct control over metabolic processes, optimizing conditions for higher yields without the limitations associated with cellular regulation. Researchers have successfully used cell-free systems for the synthesis of antibiotics and other bioactive compounds, allowing industries to rapidly test new biosynthetic pathways.

5.6 OMICS TECHNOLOGIES AND THE POTENTIAL FOR METABOLITE DIVERSIFICATION

Omics technologies—genomics, transcriptomics, proteomics, and metabolomics—have become invaluable tools in microbial research. These technologies provide a comprehensive understanding of cellular processes by analyzing the complete set of genes, mRNA, proteins, and metabolites in an organism. Omics data allow researchers to identify biosynthetic gene clusters (BGCs) responsible for secondary metabolite production, revealing potential targets for genetic modification.

Metagenomics, a subset of genomics, enables the study of genetic material from environmental samples, providing insight into secondary metabolites produced by unculturable microorganisms. By extracting DNA from soil, water, or other environments, researchers can identify new biosynthetic pathways and engineer these pathways into culturable microorganisms. This approach has led to the discovery of novel antibiotics and bioactive compounds with potential applications in medicine and agriculture.

5.7 FUTURE PROSPECTS IN MICROBIAL SYNTHESIS AND STRAIN IMPROVEMENT

The future of microbial synthesis of secondary metabolites and strain improvement is bright, driven by technological advancements in biotechnology, synthetic biology, and computational modeling.

Machine learning and artificial intelligence (AI) are likely to play an increasing role in this field, with AI algorithms capable of predicting the effects of genetic modifications on metabolite production. These predictions can guide the design of strain improvement experiments, saving time and resources.

As the need for sustainable, eco-friendly products grows, microbial synthesis of secondary metabolites has the potential to replace environmentally harmful synthetic methods.

In agriculture, for instance, microbial-derived insecticides and fungicides could reduce dependency on chemical pesticides, decreasing environmental impact and contributing to healthier ecosystems. In the pharmaceutical industry, microbial synthesis could provide a sustainable route for producing complex drug compounds, reducing costs and enabling more accessible treatments.

Another promising area is the production of personalized medicine. Advances in strain improvement techniques allow for the microbial synthesis of customized secondary metabolites, tailored to individual medical needs. For example, engineered microbes can potentially produce specific drug variants to better match a patient's genetic profile, offering a new dimension to personalized treatment.

Furthermore, researchers are exploring the use of microbial consortia, where multiple engineered strains work together to produce complex secondary metabolites. Microbial consortia mimic natural microbial communities, where microbes often rely on each other's metabolic by-products. By designing consortia with complementary metabolic capabilities, scientists can increase the yield and complexity of secondary metabolites produced.

5.8 CONCLUSION

The microbial synthesis of secondary metabolites, supported by advances in strain improvement techniques, holds great potential for various industries, including pharmaceuticals, agriculture, and biotechnology. Current trends such as co-culturing, pathway engineering, and synthetic biology have unlocked new possibilities for enhancing production efficiency and discovering novel metabolites. Omics technologies and computational tools further enable researchers to explore and manipulate microbial metabolic networks at unprecedented levels, paving the way for innovation.

Looking forward, developments in artificial intelligence, cell-free systems, and microbial consortia will continue to transform microbial synthesis, making it a key component in the bioeconomy. With these advancements, the microbial synthesis of secondary metabolites will not only meet the growing demand for bioactive compounds but will also contribute to sustainable industrial practices, reducing the ecological footprint of chemical synthesis and supporting global health and environmental goals.

CHAPTER 6

FLUX CONTROL ANALYSIS AND STOICHIOMETRIC NETWORK MODELING: CURRENT TRENDS AND FUTURE PROSPECTS

Flux control analysis (FCA) and stoichiometric network modeling (SNM) are vital tools in understanding metabolic networks within cells. These methods provide a quantitative understanding of how fluxes—the rates at which metabolites are converted through biochemical pathways—are controlled and regulated within a cell. In combination, FCA and SNM offer insights into how metabolic pathways are orchestrated, enabling scientists to design more efficient production strains, optimize metabolic pathways for biotechnological applications, and enhance the synthesis of desired metabolites.

6.1 PRINCIPLES OF FLUX CONTROL ANALYSIS AND ITS APPLICATIONS

Flux control analysis is a quantitative method used to determine the control that each enzymatic step exerts over the flux through a metabolic pathway. In metabolic networks, not all enzymes have equal influence over pathway flux; FCA identifies the "control points" that significantly impact the flow of metabolites. By evaluating flux control coefficients for enzymes within a pathway, researchers can identify critical points where interventions—such as enzyme overexpression or inhibition—could optimize metabolite production.

A primary application of FCA is in metabolic engineering, where it aids in understanding how modifications at key steps can influence overall production yield. For instance, in the synthesis of amino acids like lysine, FCA can help determine which enzymes in the biosynthetic pathway most affect production levels, allowing for targeted genetic modifications. Similarly, in biofuel production, FCA assists in identifying bottlenecks within microbial metabolic pathways, enabling engineers to overcome limitations that slow down or limit production.

In recent years, FCA has been integrated with other computational methods, such as metabolic control analysis (MCA), to offer a more comprehensive view of cellular flux control. This integration provides a deeper understanding of both linear and non-linear pathways, enhancing the accuracy and utility of FCA in complex networks. Modern FCA methods often include sensitivity analysis, which helps assess how variations in enzyme activity levels impact flux, making it a powerful tool for strain optimization and pathway design.

6.2 STOICHIOMETRIC NETWORK MODELING: FRAMEWORK AND USES

Stoichiometric network modeling focuses on the structure of metabolic networks, describing the stoichiometric relationships between metabolites and reactions without considering reaction kinetics. SNM is based on the stoichiometric matrix, a mathematical construct that represents the relationships between metabolites (rows) and reactions (columns) in a pathway. By using matrix-based models, researchers can analyze the interconnections within metabolic networks and calculate reaction rates that are consistent with the law of mass conservation.

SNM is particularly useful in identifying metabolic constraints and predicting feasible reaction rates under various conditions. In biotechnology, SNM has been employed to optimize metabolic networks for the production of biofuels, pharmaceuticals, and other valuable compounds. One of its core applications is flux balance analysis (FBA), which uses stoichiometric constraints to predict steady-state flux distributions in cells. FBA has become an essential tool in synthetic biology for predicting cellular behavior under different genetic and environmental conditions, helping scientists design microorganisms for efficient production of bio-based products.

A popular extension of SNM is elementary mode analysis (EMA), which breaks down complex networks into minimal pathways, known as elementary modes, that can operate independently of one another. EMA allows researchers to identify unique pathways for metabolite synthesis, offering a clearer picture of alternative metabolic routes within cells. This information can be invaluable for strain improvement and pathway optimization in metabolic engineering.

6.3 CURRENT TRENDS IN FLUX CONTROL ANALYSIS

Recent advancements in FCA have focused on improving its predictive power through the integration of machine learning (ML) and artificial intelligence (AI) algorithms. ML models can analyze large datasets generated by FCA, identifying patterns and suggesting potential interventions that may improve flux distribution across a pathway. This approach has proven especially useful in strain development, where identifying and optimizing key flux control points can result in increased yields of desired products, such as biofuels and pharmaceuticals.

Another current trend in FCA is multi-objective optimization, which addresses the challenge of balancing different metabolic objectives. For example, a strain engineered for biofuel production may need to maximize yield while also maintaining cell viability. FCA models that include multi-objective optimization enable researchers to balance these competing demands, leading to more robust and efficient production strains.

With advancements in high-throughput data collection, FCA can now be performed on a genome-scale level, allowing researchers to analyze entire metabolic networks rather than focusing on single pathways. Genome-scale metabolic modeling (GEM) integrates FCA with genomics, transcriptomics, and proteomics data, providing a holistic view of metabolic flux across the entire organism. GEMs are instrumental in systems biology, where they aid in predicting how genetic modifications impact global metabolism, enabling scientists to design highly efficient strains for large-scale industrial applications.

6.4 INNOVATIONS IN STOICHIOMETRIC NETWORK MODELING

SNM has also seen significant advancements, particularly in the use of constraint-based modeling techniques, which add layers of complexity to stoichiometric models by incorporating additional biological constraints. Techniques such as thermodynamic flux balance analysis (tFBA) incorporate thermodynamic properties of reactions, ensuring that predicted flux distributions are energetically feasible. This approach has improved the accuracy of SNM predictions, allowing for more realistic models that align closely with experimental results.

Another key trend in SNM is the development of kinetic-based stoichiometric models, which combine stoichiometry with reaction kinetics. These models incorporate rate laws for individual reactions, accounting for the dynamic nature of metabolic networks. While traditional SNM relies solely on stoichiometric constraints, kinetic models provide insights into time-dependent changes in metabolite concentrations, helping researchers better understand the dynamics of metabolic responses to genetic modifications or environmental changes.

With the rise of systems biology, the integration of SNM with omics data—genomics, transcriptomics, and metabolomics—has become increasingly important. This integrative approach, sometimes referred to as "multi-omics modeling," enables researchers to construct more accurate models that reflect the organism's metabolic state under various conditions. By combining SNM with omics data, scientists can make more precise predictions regarding metabolic fluxes and pathway activities, advancing strain improvement and bioprocess optimization.

6.5 FUTURE PROSPECTS IN FCA AND SNM: TOWARDS PRECISION METABOLIC ENGINEERING

The future of FCA and SNM lies in precision metabolic engineering, where scientists can tailor metabolic networks with unparalleled specificity and efficiency.

Developments in CRISPR technology, synthetic biology, and machine learning are expected to further enhance the capabilities of FCA and SNM, enabling precise control over metabolic fluxes and facilitating the design of customized metabolic pathways.

AI-powered predictive modeling will likely play a significant role in advancing FCA and SNM. With AI, researchers can analyze complex datasets generated from FCA and SNM experiments, identifying optimal genetic modifications and predicting their impact on metabolite production. These predictive models will improve the efficiency of strain development, reducing the time and cost associated with traditional trial-and-error approaches.

Synthetic biology, particularly the design of modular and synthetic pathways, is another area where FCA and SNM are likely to expand.

By combining FCA and SNM with synthetic biology tools, researchers can create entirely new pathways or redesign existing ones to maximize flux through desired pathways. These synthetic pathways could enable the production of non-native metabolites in microbial strains, opening new possibilities in pharmaceutical, agricultural, and energy sectors.

6.6 APPLICATIONS IN PERSONALIZED AND PRECISION MEDICINE

The integration of FCA and SNM in personalized medicine holds significant potential for the future. By analyzing patient-specific data, scientists can use FCA and SNM models to predict how a person's unique metabolic profile will respond to certain treatments or dietary modifications. For instance, in cancer research, FCA can be used to identify metabolic vulnerabilities in tumor cells, guiding the development of targeted therapies. Similarly, SNM models can simulate the metabolic impact of specific drugs, allowing for more personalized and effective treatments.

In addition to personalized medicine, FCA and SNM offer prospects in environmental biotechnology and sustainable bioproduction. By optimizing metabolic fluxes in microbes, researchers can enhance the efficiency of processes like carbon sequestration, bioremediation, and biofuel production. These applications align with global sustainability goals, positioning FCA and SNM as valuable tools for tackling environmental challenges.

6.7 CONCLUSION

Flux control analysis and stoichiometric network modeling are indispensable tools in modern metabolic engineering, allowing researchers to analyze, predict, and optimize metabolic networks. The current trends in FCA, including machine learning integration and genome-scale modeling, are transforming the precision and applicability of metabolic engineering. Similarly, advancements in SNM, from constraint-based modeling to multi-omics integration, have improved the ability to simulate complex metabolic networks accurately.

Looking forward, the combination of FCA, SNM, and advanced technologies like AI, CRISPR, and synthetic biology will continue to expand the potential of these approaches.

These advancements are expected to enable precision metabolic engineering for a range of applications, from sustainable biofuel production to personalized medicine, paving the way for a future where metabolic networks can be fine-tuned for specific industrial, environmental, and medical purposes. As FCA and SNM methodologies evolve, they hold immense promise in supporting both innovation and sustainability across various sectors, offering new solutions for the pressing challenges of today and tomorrow.

CHAPTER 7

CONTROL OF METABOLITE EFFLUX IN MICROBIAL CELL FACTORIES: CURRENT ADVANCES AND FUTURE PROSPECTS

Microbial cell factories are engineered microorganisms, such as bacteria, yeast, or fungi, optimized to produce valuable metabolites like biofuels, bioplastics, pharmaceuticals, and food ingredients. The effectiveness of these cell factories hinges on efficient intracellular production and, crucially, the ability to export or "efflux" these metabolites from the cell.

Efficient metabolite efflux reduces toxicity within the cell and prevents product accumulation that can slow down biosynthesis. Metabolite efflux mechanisms are increasingly engineered to meet the demands of industry, enabling high-yield production while maintaining cellular health and functionality.

7.1 THE IMPORTANCE OF METABOLITE EFFLUX IN MICROBIAL PRODUCTION

Metabolite efflux plays a central role in maintaining the balance within microbial cells, as accumulation of metabolites can hinder cell growth, cause feedback inhibition, and lower productivity. In microbial cell factories, where cells are engineered to produce high levels of target compounds, controlling the export of these products is critical.

By optimizing efflux pathways, the metabolic burden on cells can be reduced, resulting in enhanced productivity and potentially reducing the need for costly downstream processing.

Efflux control is particularly important for the production of hydrophobic compounds, acids, or other compounds that may damage cellular membranes or interfere with other cellular processes.

For instance, the production of high concentrations of fatty acids or biofuels within a microbial cell can disrupt membrane integrity, impacting cell viability. The ability to direct these compounds out of the cell at the optimal rate is thus essential to maintain a productive microbial cell factory.

7.2 MECHANISMS OF METABOLITE EFFLUX IN MICROORGANISMS

In microorganisms, efflux systems are primarily composed of transporter proteins that actively pump or passively diffuse metabolites across the cellular membrane. These transporters can be broadly categorized into primary and secondary transport systems. Primary transporters use energy derived from ATP hydrolysis to actively transport molecules across the membrane, while secondary transporters rely on proton or ion gradients to facilitate transport.

Key classes of efflux transporters include ATP-binding cassette (ABC) transporters, which actively pump metabolites out of the cell using ATP; major facilitator superfamily (MFS) transporters, which facilitate transport based on ion gradients; and resistance-nodulation-cell division (RND) transporters, often found in bacteria and responsible for exporting a wide range of compounds, including antibiotics. Each type of transporter has specific advantages and disadvantages, depending on the nature of the metabolite and the desired efflux characteristics.

Recent research has explored ways to engineer these transporters to enhance metabolite export in microbial cell factories. The ability to modulate the expression of these transporters, or even redesign them for higher specificity and efficiency, is central to improving the overall productivity of microbial cell factories.

7.3 ADVANCES IN GENETIC ENGINEERING OF EFFLUX SYSTEMS

One major advance in efflux control is the use of genetic engineering to enhance or introduce transporters specific to the target metabolite. Genetic tools such as CRISPR-Cas9 have enabled precise control over transporter genes, allowing researchers to either upregulate native transporters or introduce heterologous (foreign) transporters from other organisms. This approach has been successful in enhancing the efflux of a variety of bio-products, including biofuels, organic acids, and therapeutic compounds.

For example, in the production of isoprenoids—a class of compounds used in biofuels, fragrances, and pharmaceuticals—engineers have introduced heterologous transporters from plants or other microbes to enhance export in *Escherichia coli.* By incorporating these plant-derived transporters, isoprenoid efflux is optimized, reducing the toxic effects on cells and enhancing productivity. Similarly, in lactic acid production, genetic modifications to express more effective lactate transporters in *Lactobacillus* species have improved acid tolerance and production rates. Another promising development is the engineering of inducible efflux systems, where transporters are activated in response to metabolite accumulation. These systems, often controlled by metabolite-sensing promoters, allow transporters to be expressed only when needed, reducing the energy burden on cells during the early stages of growth and ensuring optimal timing for efflux.

7.4 SYNTHETIC BIOLOGY AND EFFLUX OPTIMIZATION

Synthetic biology has revolutionized the control of metabolite efflux in microbial cell factories. Synthetic biology tools enable the design of custom transporters and efflux pathways to match the unique needs of each product. This field has led to the creation of synthetic transporter genes tailored to export specific metabolites, or even to export multiple compounds simultaneously.

With advances in protein engineering, researchers can now design transporters with altered specificity, affinity, and energy requirements. For instance, computational models can predict the structure and function of engineered transporters, enabling precise modifications that enhance the transport of desired compounds. One synthetic biology approach uses directed evolution to generate transporters that can handle high concentrations of product without losing efficiency. By creating libraries of transporter variants and selecting those with improved performance, scientists can evolve transporters to meet specific efflux needs.

Another synthetic biology tool is the design of synthetic biosensors that detect metabolite levels and regulate transporter expression accordingly. These biosensors can be linked to transcription factors that activate transporter genes when metabolite concentrations exceed a threshold, ensuring that export occurs only when needed. Such biosensors have been developed for several microbial products, including bioethanol, lactic acid, and secondary metabolites, providing a level of efflux control that is responsive to metabolic conditions.

Understanding the Synthetic Biology

Synthetic biology (SynBio) is a multidisciplinary field of science that focuses on living systems and organisms, and it applies engineering principles to develop new biological parts, devices, and systems or to redesign existing systems found in nature. It is a branch of science that encompasses a broad range of methodologies from various disciplines, such as biochemistry, biotechnology, biomaterials, material science/engineering, genetic engineering, molecular biology, molecular engineering, systems biology, membrane science, biophysics, chemical and biological engineering, electrical and computer engineering, control engineering and evolutionary biology.

It includes designing and constructing biological modules, biological systems, and biological machines, or re-designing existing biological systems for useful purposes. Additionally, it is the branch of science that focuses on the new abilities of engineering into existing organisms to redesign them for useful purposes. In order to produce predictable and robust systems with novel functionalities that do not already exist in nature, it is also necessary to apply the engineering paradigm of systems design to biological systems. According to the European Commission, this possibly involves a molecular assembler based on biomolecular systems such as the ribosome.

History

- **1910:** First identifiable use of the term *synthetic biology* in Stéphane Leduc's publication *Théorie physico-chimique de la vie et générations spontanées*. He also noted this term in another publication, *La Biologie Synthétique* in 1912.
- **1944:** Canadian-American scientist Oswald Avery shows that DNA is the material of which genes and chromosomes are made. This becomes the bedrock on which all subsequent genetic research is built.
- **1953:** Francis Crick and James Watson publish the structure of the DNA in *Nature*.
- **1961:** Jacob and Monod postulate cellular regulation by molecular networks from their study of the *lac* operon in *E. coli* and envisioned the ability to assemble new systems from molecular components.
- **1973:** First molecular cloning and amplification of DNA in a plasmid is published in *P.N.A.S.* by Cohen, Boyer et al. constituting the dawn of synthetic biology.
- **1978:** Arber, Nathans and Smith win the Nobel Prize in Physiology or Medicine for the discovery of restriction enzymes, leading Szybalski to offer an editorial comment in the journal *Gene*:
  - The work on restriction nucleases not only permits us easily to construct recombinant DNA molecules and to analyze individual genes, but also has led us into the new era of synthetic biology where not only existing genes are described and analyzed but also new gene arrangements can be constructed and evaluated.
- **1988:** First DNA amplification by the polymerase chain reaction (PCR) using a thermostable DNA polymerase is published in *Science* by Mullis et al. This obviated adding new DNA polymerase after each PCR cycle, thus greatly simplifying DNA mutagenesis and assembly.
- **2000:** Two papers in *Nature* report synthetic biological circuits, a genetic toggle switch and a biological clock, by combining genes within *E. coli* cells.
- **2003:** The most widely used standardized DNA parts, BioBrick plasmids, are invented by Tom Knight. These parts will become central to the International Genetically Engineered Machine (iGEM) competition founded at MIT in the following year.

- **2003:** Researchers engineer an artemisinin precursor pathway in *E. coli.*
- **2004:** First international conference for synthetic biology, Synthetic Biology 1.0 (SB1.0) is held at MIT.
- **2005:** Researchers develop a light-sensing circuit in *E. coli.* Another group designs circuits capable of multicellular pattern formation.
- **2006:** Researchers engineer a synthetic circuit that promotes bacterial invasion of tumour cells.
- **2010:** Researchers publish in *Science* the first synthetic bacterial genome, called *M. mycoides* JCVI-syn1.0. The genome is made from chemically-synthesized DNA using yeast recombination.
- **2011:** Functional synthetic chromosome arms are engineered in yeast.
- **2012:** Charpentier and Doudna labs publish in *Science* the programming of CRISPR-Cas9 bacterial immunity for targeting DNA cleavage. This technology greatly simplified and expanded eukaryotic gene editing.
- **2019:** Scientists at ETH Zurich report the creation of the first bacterial genome, named *Caulobacter ethensis-2.0,* made entirely by a computer, although a related viable form of *C. ethensis-2.0* does not yet exist.
- **2019:** Researchers report the production of a new synthetic (possibly artificial) form of viable life, a variant of the bacteria *Escherichia coli,* by reducing the natural number of 64 codons in the bacterial genome to 59 codons instead, in order to encode 20 amino acids.
- **2020:** Scientists created the first xenobot, a programmable synthetic organism derived from frog cells and designed by AI.
- **2021:** Scientists reported that xenobots are able to self-replicate by gathering loose cells in the environment and then forming new xenobots.

Perspectives

It is a field whose scope is expanding in terms of systems integration, engineered organisms, and practical findings.

Engineers view biology as *technology* (in other words, a given system includes *biotechnology* or its *biological engineering*). Synthetic biology includes the broad redefinition and expansion of biotechnology, with the ultimate goal of being able to design and build engineered live biological systems that process information, manipulate chemicals, fabricate materials and structures, produce energy, provide food, and maintain and enhance human health, as well as advance fundamental knowledge of biological systems and our environment.

Researchers and companies working in synthetic biology are using nature's power to solve issues in agriculture, manufacturing, and medicine.

Due to more powerful genetic engineering capabilities and decreased DNA synthesis and sequencing costs, the field of synthetic biology is rapidly growing. In 2016, more than 350 companies across 40 countries were actively engaged in synthetic biology applications; all these companies had an estimated net worth of $3.9 billion in the global market. Synthetic biology currently has no generally accepted definition. Here are a few examples:

It is the science of emerging genetic and physical engineering to produce new (and, therefore, synthetic) life forms. To develop organisms with novel or enhanced characteristics, this emerging field of study combines biology, engineering, and related disciplines' knowledge and techniques to design chemically synthesised DNA.

Biomolecular engineering includes approaches that aim to create a toolkit of functional units that can be introduced to present new technological functions in living cells. Genetic engineering includes approaches to construct synthetic chromosomes or minimal organisms like *Mycoplasma laboratorium.*

Biomolecular design refers to the general idea of de novo design and additive combination of biomolecular components. Each of these approaches shares a similar task: to develop a more synthetic entity at a higher level of complexity by inventively manipulating a simpler part at the preceding level. Optimizing these exogenous pathways in unnatural systems takes iterative fine-tuning of the individual biomolecular components to select the highest concentrations of the desired product. On the other hand, "re-writers" are synthetic biologists interested in testing the irreducibility of biological systems. Due to the complexity of natural biological systems, it would be simpler to rebuild the natural systems of interest from the ground up; to provide engineered surrogates that are easier to comprehend, control and manipulate. Re-writers draw inspiration from refactoring, a process sometimes used to improve computer software.

Categories

Bioengineering, synthetic genomics, protocell synthetic biology, unconventional molecular biology, and in silico techniques are the five categories of synthetic biology. It is necessary to review the distinctions and analogies between the categories of synthetic biology for its social and ethical assessment, to distinguish between issues affecting the whole field and particular to a specific one.

Bioengineering

The subfield of bioengineering concentrates on creating novel metabolic and regulatory pathways, and is currently the one that likely draws the attention of most researchers and funding. It is primarily motivated by the desire to establish biotechnology as a legitimate engineering discipline. When referring to this area of synthetic biology, the word "bioengineering" should not be confused with "traditional genetic engineering", which involves introducing a single transgene into the intended organism. Bioengineers adapted synthetic biology to provide a substantially more integrated perspective on how to alter organisms or metabolic systems.

A typical example of single-gene genetic engineering is the insertion of the human insulin gene into bacteria to create transgenic proteins. The creation of whole new signalling pathways, containing numerous genes and regulatory components (such as an oscillator circuit to initiate the periodic production of green fluorescent protein (GFP) in mammalian cells), is known as bioengineering as part of synthetic biology.

By utilising simplified and abstracted metabolic and regulatory modules as well as other standardized parts that may be freely combined to create new pathways or creatures, bioengineering aims to create innovative biological systems. In addition to creating infinite opportunities for novel applications, this strategy is anticipated to make bioengineering more predictable and controllable than traditional biotechnology.

Synthetic Genomics

The formation of animals with a chemically manufactured (minimal) genome is another facet of synthetic biology that is highlighted by synthetic genomics. This area of synthetic biology has been made possible by ongoing advancements in DNA synthesis technology, which now makes it feasible to produce DNA molecules with thousands of base pairs at a reasonable cost. The goal is to combine these molecules into complete genomes and transplant them into living cells, replacing the host cell's genome and reprogramming its metabolism to perform different functions.

Scientists have previously demonstrated the potential of this approach by creating infectious viruses by synthesising the genomes of multiple viruses. These significant advances in science and technology triggered the initial public concerns concerning the risks associated with this technology.

A simple genome might also work as a "chassis genome" that could be enlarged quickly by gene inclusion created for particular tasks. Such "chassis creatures" would be more suited for the insertion of new functions than wild organisms since they would have fewer biological pathways that could potentially conflict with the new functionalities in addition to having specific insertion sites. Synthetic genomics strives to create creatures with novel "architectures," much like the bioengineering method. It adopts an integrative or holistic perspective of the organism. In this case, the objective is the creation of chassis genomes based on necessary genes and other required DNA sequences rather than the design of metabolic or regulatory pathways based on abstract criteria.

Protocell Synthetic Biology

The in vitro generation of synthetic cells is the protocell branch of synthetic biology. Lipid vesicles, which have all the necessary components to function as a complete system, can be used to create these artificial cells. In the end, these synthetic cells should meet the requirements for being deemed alive, namely the capacity for self-replication, self-maintenance, and evolution. The protocell technique has this as its end aim, however there are other intermediary steps that fall short of meeting all the criteria for a living cell. In order to carry out a specific function, these lipid vesicles contain cell extracts or more specific sets of biological macromolecules and complex structures, such as enzymes, nucleic acids, or ribosomes. For instance, liposomes may carry out particular polymerase chain reactions or synthesise a particular protein.

Protocell synthetic biology takes artificial life one step closer to reality by eventually synthesizing not only the genome but also every component of the cell in vitro, as opposed to the synthetic genomics approach, which relies on coercing a natural cell to carry out the instructions encoded by the introduced synthetic genome. Synthetic biologists in this field view their work as basic study into the conditions necessary for life to exist and its origin more than in any of the other techniques. The protocell technique, however, also lends itself well to applications; similar to other synthetic biology byproducts, protocells could be employed for the manufacture of biopolymers and medicines.

Unconventional Molecular Biology

The objective of the "unnatural molecular biology" strategy is to create new varieties of life that are based on a different kind of molecular biology, such as new types of nucleic acids or a new genetic code. The creation of new types of nucleotides that can be built into unique nucleic acids could be accomplished by changing certain DNA or RNA constituents, such as the bases or the backbone sugars.

The normal genetic code is being altered by inserting quadruplet codons or changing some codons to encode new amino acids, which would subsequently permit the use of non-natural amino acids with unique features in protein production. It is a scientific and technological problem to adjust the enzymatic machinery of the cell for both approaches.

A new sort of life would be formed by organisms with a genome built on synthetic nucleic acids or on a totally new coding system for synthetic amino acids. This new style of life would have some benefits but also some new dangers. On release into the environment, there would be no horizontal gene transfer or outcrossing of genes with natural species. Furthermore, these kinds of synthetic organisms might be created to require non-natural materials for protein or nucleic acid synthesis, rendering them unable to thrive in the wild if they accidentally escaped.

On the other hand, if these organisms ultimately were able to survive outside of controlled space, they might have a particular benefit over natural organisms because they would be resistant to predatory living organisms or natural viruses, that could lead to an unmanaged spread of the synthetic organisms.

In Silico Technique

Synthetic biology in silico and the various strategies are interconnected. The development of complex designs, whether they are metabolic pathways, fundamental cellular processes, or chassis genomes, is one of the major difficulties faced by the four synthetic-biology methods outlined above. Because of this, synthetic biology has a robust in silico branch, similar to systems biology, that aims to create computational models for the design of common biological components or synthetic circuits, which are essentially simulations of synthetic organisms.

The practical application of simulations and models through bioengineering or other fields of synthetic biology is the long-term goal of in silico synthetic biology. Many of the computational simulations of synthetic organisms up to this point possess little to no direct analogy to living things. Due to this, in silico synthetic biology is regarded as a separate group in this article.

It is sensible to integrate the five areas under the umbrella of synthetic biology as an unified area of study.

Even though they focus on various facets of life, such as metabolic regulation, essential elements, or biochemical makeup, these five strategies all work toward the same end: creating new types of living organisms. Additionally, the varied methodologies begin with numerous methodological approaches, which leads to the diversity of synthetic biology approaches.

Synthetic biology is an interdisciplinary field that draws from and is inspired by many different scientific disciplines, not one single field or technique. Synthetic biologists all have the same underlying objective of designing and producing new forms of life, despite the fact that they may employ various methodologies, techniques, and research instruments.

Any evaluation of synthetic biology, whether it examines ethical, legal, or safety considerations, must take into account the fact that while some questions, risks, and issues are unique to each technique, in other circumstances, synthetic biology as a whole must be taken into consideration.

Four Engineering Approaches

Synthetic biology has traditionally been divided into four different engineering approaches: top down, parallel, orthogonal and bottom up. To replicate emergent behaviours from natural biology and build artificial life, unnatural chemicals are used. The other looks for interchangeable components from biological systems to put together and create systems that do not work naturally. In either case, a synthetic objective compels researchers to venture into new area in order to engage and resolve issues that cannot be readily resolved by analysis. Due to this, new paradigms are driven to arise in ways that analysis cannot easily do. In addition to equipments that oscillate, creep, and play tic-tac-toe, synthetic biology has produced diagnostic instruments that enhance the treatment of patients with infectious diseases.

Top-Down Approach

It involves using metabolic and genetic engineering techniques to impart new functions to living cells. By comparing universal genes and eliminating non-essential ones to create a basic genome, this method seeks to lessen the complexity of existing cells. These initiatives are founded on the hypothesis of a single genesis for cellular life, the so-called Last Universal Common Ancestor, which supports the presence of a universal minimal genome that gave rise to all living things. Recent studies, however, raise the possibility that the eukaryotic and prokaryotic cells that make up the tree of life may have evolved from a group of primordial cells rather than from a single cell. As a result, even while the Holy Grail-like pursuit of the "minimum genome" has grown elusive, cutting out a number of non-essential functions impairs an organism's fitness and leads to "fragile" genomes.

Bottom-Up Approach

This approach involves creating new biological systems in vitro by bringing together 'non-living' biomolecular components, often with the aim of constructing an artificial cell. Reproduction, replication, and assembly are three crucial self-organizational principles that are taken into account in order to accomplish this. Cells, which are made up of a container and a metabolism, are considered "hardware" in the definition of reproduction, whereas replication occurs when a system duplicates a perfect copy of itself, as in the case of DNA, which is considered "software." When vesicles or containers (such as Oparin's coacervates) formed of tiny droplets of molecules that are organic like lipids or liposomes, membrane-like structures comprising phospholipids, aggregate, assembly occur.

The study of protocells exists along with other in vitro synthetic biology initiatives that seek to produce minimum cells, metabolic pathways, or "never-born proteins" as well as to mimic physiological functions including cell division and growth. The in vitro enhancement of synthetic pathways does have the potential to have an effect on some other synthetic biology sectors, including metabolic engineering, despite the fact that it no longer classified as synthetic biology research. This research, which is primarily essential, deserves proper recognition as synthetic biology research.

Parallel Approach

Parallel engineering is also known as bioengineering. The basic genetic code is the foundation for parallel engineering research, which uses conventional biomolecules like nucleic acids and the 20 amino acids to construct biological systems. For a variety of applications in biocomputing, bioenergy, biofuels, bioremediation, optogenetics, and medicine, it involves the standardisation of DNA components, engineering of switches, biosensors, genetic circuits, logic gates, and cellular communication operators. For directing the expression of two or more genes and/ or proteins, the majority of these applications often rely on the use of one or more vectors (or plasmids). Small, circular, double-strand DNA units known as plasmids, which are primarily found in prokaryotic but can also occasionally be detected in eukaryotic cells, may replicate autonomously of chromosomal DNA.

Orthogonal Approach

It is also known as perpendicular engineering. This strategy, also referred to as "chemical synthetic biology," principally seeks to alter or enlarge the genetic codes of living systems utilising artificial DNA bases and/or amino acids. This subfield is also connected to xenobiology, a newly developed field that combines systems chemistry, synthetic biology, exobiology, and research into the origins of life. In recent decades, researchers have created compounds that are structurally similar to the DNA canonical bases to see if those "alien" or xeno (XNA) molecules may be employed as genetic information carriers. Similar to this, noncanonical moieties have taken the place of the DNA sugar (deoxyribose).

In order to express information other than the 20 conventional amino acids of proteins, the genetic code can be altered or enlarged. One method involves incorporating a specified unnatural, noncanonical, or xeno amino acid (XAA) into one or more proteins at one or more precise places using orthogonal enzymes and a transfer RNA adaptor from an other organism.

By using "directed evolution," which entails repeated cycles of gene mutagenesis (genotypic diversity production), screening or selection (of a specific phenotypic trait), and amplification of a better variant for the following iterative round, orthogonal enzymes are produced Numerous XAAs have been effectively incorporated into proteins in more complex creatures like worms and flies as well as in bacteria, yeast, and human cell lines.

As a result of canonical DNA sequence changes, directed evolution also enables the development of orthogonal ribosomes, which make it easier to incorporate XAAs into proteins or create "mirror life," or biological systems that contain biomolecules made up of enantiomers with different chiral orientations.

Enabling Technologies

Several novel enabling technologies were critical to the success of synthetic biology. Concepts include standardization of biological parts and hierarchical abstraction to permit using those parts in synthetic systems. DNA serves as the guide for how biological processes should function, like the score to a complex symphony of life. Our ability to comprehend and design biological systems has undergone significant modifications as a result of developments in the previous few decades in both reading (sequencing) and writing (synthesis) DNA sequences.

These developments have produced ground-breaking techniques for designing, assembling, and modifying DNA-encoded genes, materials, circuits, and metabolic pathways, enabling an ever-increasing amount of control over biological systems and even entire organisms. Basic technologies include reading and writing DNA (sequencing and fabrication). Measurements under multiple conditions are needed for accurate modeling and computer-aided design (CAD).

DNA and Gene Synthesis

Driven by dramatic decreases in costs of oligonucleotide ("oligos") synthesis and the advent of PCR, the sizes of DNA constructions from oligos have increased to the genomic level. In 2000, researchers reported synthesis of the 9.6 kbp (kilo bp) Hepatitis C virus genome from chemically synthesized 60 to 80-mers. In 2002, researchers at Stony Brook University succeeded in synthesizing the 7741 bp poliovirus genome from its published sequence, producing the second synthetic genome, spanning two years.

In 2003, the 5386 bp genome of the bacteriophage Phi X 174 was assembled in about two weeks. In 2006, the same team, at the J. Craig Venter Institute, constructed and patented a synthetic genome of a novel minimal bacterium, *Mycoplasma laboratorium* and were working on getting it functioning in a living cell.

In 2007, it was reported that several companies were offering synthesis of genetic sequences up to 2000 base pairs (bp) long, for a price of about $1 per bp and a turnaround time of less than two weeks.

Oligonucleotides harvested from a photolithographic- or inkjet-manufactured DNA chip combined with PCR and DNA mismatch error-correction allows inexpensive large-scale changes of codons in genetic systems to improve gene expression or incorporate novel amino-acids. This favors a synthesis-from-scratch approach.

Additionally, the CRISPR/Cas system has emerged as a promising technique for gene editing. It was described as "the most important innovation in the synthetic biology space in nearly 30 years". While other methods take months or years to edit gene sequences, CRISPR speeds that time up to weeks. Due to its ease of use and accessibility, however, it has raised ethical concerns, especially surrounding its use in biohacking.

Sequencing

DNA sequencing determines the order of nucleotide bases in a DNA molecule. Synthetic biologists use DNA sequencing in their work in several ways. First, large-scale genome sequencing efforts continue to provide information on naturally occurring organisms. This information provides a rich substrate from which synthetic biologists can construct parts and devices. Second, sequencing can verify that the fabricated system is as intended. Third, fast, cheap, and reliable sequencing can facilitate rapid detection and identification of synthetic systems and organisms.

Modularity

This is the ability of a system or component to operate without reference to its context. The most used standardized DNA parts are BioBrick plasmids, invented by Tom Knight in 2003. Biobricks are stored at the Registry of Standard Biological Parts in Cambridge, Massachusetts. The BioBrick standard has been used by tens of thousands of students worldwide in the international Genetically Engineered Machine (iGEM) competition. BioBrick Assembly Standard 10 promotes modularity by allowing BioBrick coding sequences to be spliced out and exchanged using restriction enzymes EcoRI or XbaI (BioBrick prefix) and SpeI and PstI (BioBrick suffix).

Sequence overlap between two genetic elements (genes or coding sequences), called overlapping genes, can prevent their individual manipulation. To increase genome modularity, the practice of genome refactoring or improving "the internal structure of an existing system for future use, while simultaneously maintaining external system function" has been adopted across synthetic biology disciplines. Some notable examples of refactoring including the nitrogen fixation cluster and type III secretion system along with bacteriophages T7 and ΦX174.

While DNA is most important for information storage, a large fraction of the cell's activities are carried out by proteins. Tools can send proteins to specific regions of the cell and to link different proteins together. The interaction strength between protein partners should be tunable between a lifetime of seconds (desirable for dynamic signaling events) up to an irreversible interaction (desirable for device stability or resilient to harsh conditions). Interactions such as coiled coils, SH3 domain-peptide binding or SpyTag/SpyCatcher offer such control. In addition, it is necessary to regulate protein-protein interactions in cells, such as with light (using light-oxygen-voltage-sensing domains) or cell-permeable small molecules by chemically induced dimerization.

In a living cell, molecular motifs are embedded in a bigger network with upstream and downstream components. These components may alter the signaling capability of the modeling module. In the case of ultrasensitive modules, the sensitivity contribution of a module can differ from the sensitivity that the module sustains in isolation.

Modeling

Models inform the design of engineered biological systems by better predicting system behavior prior to fabrication. Synthetic biology benefits from better models of how biological molecules bind substrates and catalyze reactions, how DNA encodes the information needed to specify the cell and how multi-component integrated systems behave.

Multiscale models of gene regulatory networks focus on synthetic biology applications. Simulations can model all biomolecular interactions in transcription, translation, regulation and induction of gene regulatory networks.

Only extensive modelling can enable the exploration of dynamic gene expression in a form suitable for research and design due to the numerous involved species and the intricacy of their relationships. Dynamic simulations of the entire biomolecular interconnection involved in regulation, transport, transcription, induction, and translation enable the molecular level detailing of designs. As opposed to modelling artificial networks a posteriori, this is contrasted.

Microfluidics

Microfluidics, in particular droplet microfluidics, is an emerging tool used to construct new components, and to analyze and characterize them. It is widely employed in screening assays.

Synthetic Transcription Factors

Studies have considered the components of the DNA transcription mechanism. One desire of scientists creating synthetic biological circuits is to be able to control the transcription of synthetic DNA in unicellular organisms (prokaryotes) and in multicellular organisms (eukaryotes). One study tested the adjustability of synthetic transcription factors (sTFs) in areas of transcription output and cooperative ability among multiple transcription factor complexes. Researchers were able to mutate functional regions called zinc fingers, the DNA specific component of sTFs, to decrease their affinity for specific operator DNA sequence sites, and thus decrease the associated site-specific activity of the sTF (usually transcriptional regulation). They further used the zinc fingers as components of complex-forming sTFs, which are the eukaryotic translation mechanisms.

Applications

Synthetic biology initiatives frequently aim to redesign organisms so that they can create a material, such as a drug or fuel, or acquire a new function, such as the ability to sense something in the environment. Examples of what researchers are creating using synthetic biology include:

- Utilizing microorganisms for bioremediation to remove contaminants from our water, soil, and air.
- Production of complex natural products that are usually extracted from plants but cannot be obtained in sufficient amounts, e.g. drugs of natural origin, such as artemisinin and paclitaxel.
- Beta-carotene, a substance typically associated with carrots that prevents vitamin A deficiency, is produced by rice that has been modified. Every year, between 250,000 and 500,000 children lose their vision due to vitamin A deficiency, which also significantly raises their chance of dying from infectious infections.
- As a sustainable and environmentally benign alternative to the fresh roses that perfumers use to create expensive smells, yeast has been created to produce rose oil.

Biosensors

A biosensor refers to an engineered organism, usually a bacterium, that is capable of reporting some ambient phenomenon such as the presence of heavy metals or toxins. One such system is the Lux operon of *Aliivibrio fischeri,* which codes for the enzyme that is the source of bacterial bioluminescence, and can be placed after a respondent promoter to express the luminescence genes in response to a specific environmental stimulus. One such sensor created, consisted of a bioluminescent bacterial coating on a photosensitive computer chip to detect certain petroleum pollutants. When the bacteria sense the pollutant, they luminesce. Another example of a similar mechanism is the detection of landmines by an engineered *E.coli* reporter strain capable of detecting TNT and its main degradation product DNT, and consequently producing a green fluorescent protein (GFP).

Modified organisms can sense environmental signals and send output signals that can be detected and serve diagnostic purposes. Microbe cohorts have been used. Biosensors could also be used to detect pathogenic signatures—such as of SARS-CoV-2—and can be wearable. For the purpose of detecting and reacting to various and temporary environmental factors, cells have developed a wide range of regulatory circuits, ranging from transcriptional to post-translational. These circuits are made up of transducer modules that filter the signals and activate a biological response, as well as carefully designed sensitive sections that attach analytes and regulate signal-detection thresholds. Modularity and selectivity are programmed to biosensor circuits at the transcriptional, translational, and post-translational levels, to achieve the delicate balancing of the two basic sensing modules.

Food and Drink

Cellular agriculture focuses on the production of agricultural products from cell cultures using a combination of biotechnology, tissue engineering, molecular biology, and synthetic biology to create and design new methods of producing proteins, fats, and tissues that would otherwise come from traditional agriculture. Most of the industry is focused on animal products such as meat, milk, and eggs, produced in cell culture rather than raising and slaughtering farmed livestock which is associated with substantial global problems of detrimental environmental impacts (e.g. of meat production), animal welfare, food security and human health. Cellular agriculture is a field of the biobased economy. The most well known cellular agriculture concept is cultured meat.

However, not all synthetic nutrition products are animal food products – for instance, as of 2021, there are also products of synthetic coffee that are reported to be close to commercialization. Similar fields of research and production based on synthetic biology that can be used for the production of food and drink are:

- Genetically engineered microbial food cultures (e.g. for solar-energy-based protein powder).
- Cell-free artificial synthesis.

Materials

Photosynthetic microbial cells have been used as a step to synthetic production of spider silk.

Biological Computers

A biological computer refers to an engineered biological system that can perform computer-like operations, which is a dominant paradigm in synthetic biology. Researchers built and characterized a variety of logic gates in a number of organisms, and demonstrated both analog and digital computation in living cells. They demonstrated that bacteria can be engineered to perform both analog and/or digital computation. In 2007, in human cells, research demonstrated a universal logic evaluator that operates in mammalian cells. Subsequently, researchers utilized this paradigm to demonstrate a proof-of-concept therapy that uses biological digital computation to detect and kill human cancer cells in 2011. In 2016, another group of researchers demonstrated that principles of computer engineering can be used to automate digital circuit design in bacterial cells. In 2017, researchers demonstrated the 'Boolean logic and arithmetic through DNA excision' (BLADE) system to engineer digital computation in human cells. In 2019, researchers implemented a perceptron in biological systems opening the way for machine learning in these systems.

Cell Transformation

Cells use interacting genes and proteins, which are called gene circuits, to implement diverse function, such as responding to environmental signals, decision making and communication. Three key components are involved: DNA, RNA and Synthetic biologist designed gene circuits that can control gene expression from several levels including transcriptional, post-transcriptional and translational levels.

Traditional metabolic engineering has been bolstered by the introduction of combinations of foreign genes and optimization by directed evolution. This includes engineering *E. coli* and yeast for commercial production of a precursor of the antimalarial drug, Artemisinin.

Entire organisms have yet to be created from scratch, although living cells can be transformed with new DNA. Several ways allow constructing synthetic DNA components and even entire synthetic genomes, but once the desired genetic code is obtained, it is integrated into a living cell that is expected to manifest the desired new capabilities or phenotypes while growing and thriving. Cell transformation is used to create biological circuits, which can be manipulated to yield desired outputs.

By integrating synthetic biology with materials science, it would be possible to use cells as microscopic molecular foundries to produce materials whose properties were genetically encoded. Re-engineering has produced Curli fibers, the amyloid component of extracellular material of biofilms, as a platform for programmable nanomaterial. These nanofibers were genetically constructed for specific functions, including adhesion to substrates, nanoparticle templating and protein immobilization.

Designed Proteins

Natural proteins can be engineered, for example, by directed evolution, novel protein structures that match or improve on the functionality of existing proteins can be produced. One group generated a helix bundle that was capable of binding oxygen with similar properties as hemoglobin, yet did not bind carbon monoxide.

A similar protein structure was generated to support a variety of oxidoreductase activities while another formed a structurally and sequentially novel ATPase. Another group generated a family of G-protein coupled receptors that could be activated by the inert small molecule clozapine N-oxide but insensitive to the native ligand, acetylcholine; these receptors are known as DREADDs.

Novel functionalities or protein specificity can also be engineered using computational approaches. One study was able to use two different computational methods: a bioinformatics and molecular modeling method to mine sequence databases, and a computational enzyme design method to reprogram enzyme specificity. Both methods resulted in designed enzymes with greater than 100 fold specificity for production of longer chain alcohols from sugar.

Another common investigation is expansion of the natural set of 20 amino acids. Excluding stop codons, 61 codons have been identified, but only 20 amino acids are coded generally in all organisms.

Certain codons are engineered to code for alternative amino acids including: nonstandard amino acids such as O-methyl tyrosine; or exogenous amino acids such as 4-fluorophenylalanine. Typically, these projects make use of re-coded nonsense suppressor tRNA-Aminoacyl tRNA synthetase pairs from other organisms, though in most cases substantial engineering is required.

Other researchers investigated protein structure and function by reducing the normal set of 20 amino acids. Limited protein sequence libraries are made by generating proteins where groups of amino acids may be replaced by a single amino acid.

For instance, several non-polar amino acids within a protein can all be replaced with a single non-polar amino acid. One project demonstrated that an engineered version of Chorismate mutase still had catalytic activity when only nine amino acids were used.

Researchers and companies practice synthetic biology to synthesize industrial enzymes with high activity, optimal yields and effectiveness. These synthesized enzymes aim to improve products such as detergents and lactose-free dairy products, as well as make them more cost effective. The improvements of metabolic engineering by synthetic biology is an example of a biotechnological technique utilized in industry to discover pharmaceuticals and fermentive chemicals. Synthetic biology may investigate modular pathway systems in biochemical production and increase yields of metabolic production. Artificial enzymatic activity and subsequent effects on metabolic reaction rates and yields may develop "efficient new strategies for improving cellular properties ... for industrially important biochemical production".

Designed Nucleic Acid Systems

Scientists can encode digital information onto a single strand of synthetic DNA. In 2012, George M. Church encoded one of his books about synthetic biology in DNA. The 5.3 Mb of data was more than 1000 times greater than the previous largest amount of information to be stored in synthesized DNA. A similar project encoded the complete sonnets of William Shakespeare in DNA. More generally, algorithms such as NUPACK, ViennaRNA, Ribosome Binding Site Calculator, Cello, and Non-Repetitive Parts Calculator enables the design of new genetic systems.

Many technologies have been developed for incorporating unnatural nucleotides and amino acids into nucleic acids and proteins, both in vitro and in vivo. For example, in May 2014, researchers announced that they had successfully introduced two new artificial nucleotides into bacterial DNA. By including individual artificial nucleotides in the culture media, they were able to exchange the bacteria 24 times; they did not generate mRNA or proteins able to use the artificial nucleotides.

Space Exploration

Synthetic biology raised NASA's interest as it could help to produce resources for astronauts from a restricted portfolio of compounds sent from Earth. On Mars, in particular, synthetic biology could lead to production processes based on local resources, making it a powerful tool in the development of occupied outposts with less dependence on Earth. Work has gone into developing plant strains that are able to cope with the harsh Martian environment, using similar techniques to those employed to increase resilience to certain environmental factors in agricultural crops.

Synthetic Life

One important topic in synthetic biology is *synthetic life*, that is concerned with hypothetical organisms created *in vitro* from biomolecules and/or chemical analogues thereof. Synthetic life experiments attempt to either probe the origins of life, study some of the properties of life, or more ambitiously to recreate life from non-living (abiotic) components. Synthetic life biology attempts to create living organisms capable of carrying out important functions, from manufacturing pharmaceuticals to detoxifying polluted land and water. In medicine, it offers prospects of using designer biological parts as a starting point for new classes of therapies and diagnostic tools.

A living "artificial cell" has been defined as a completely synthetic cell that can capture energy, maintain ion gradients, contain macromolecules as well as store information and have the ability to mutate. Nobody has been able to create such a cell. A completely synthetic bacterial chromosome was produced in 2010 by Craig Venter, and his team introduced it to genomically emptied bacterial host cells. The host cells were able to grow and replicate. The *Mycoplasma laboratorium* is the only living organism with completely engineered genome.

The first living organism with 'artificial' expanded DNA code was presented in 2014; the team used *E. coli* that had its genome extracted and replaced with a chromosome with an expanded genetic code. The nucleosides added are d5SICS and dNaM.

In May 2019, in a milestone effort, researchers reported the creation of a new synthetic (possibly artificial) form of viable life, a variant of the bacteria *Escherichia coli*, by reducing the natural number of 64 codons in the bacterial genome to 59 codons instead, in order to encode 20 amino acids.

In 2017, the international Build-a-Cell large-scale open-source research collaboration for the construction of synthetic living cells was started, followed by national synthetic cell organizations in several countries, including FabriCell, MaxSynBio and BaSyC. The European synthetic cell efforts were unified in 2019 as SynCellEU initiative.

In 2023, researchers were able to create the first synthetically made human embryos derived from stem cells.

Drug Delivery Platforms

In therapeutics, synthetic biology has achieved significant advancements in altering and simplifying the therapeutics scope in a relatively short period of time. In fact, new therapeutic platforms, from the discovery of disease mechanisms and drug targets to the manufacture and transport of small molecules, are made possible by the logical and model-guided design construction of biological components.

Synthetic biology devices have been designed to act as therapies in therapeutic treatment. It is possible to control complete created viruses and organisms to target particular pathogens and diseased pathways. Thus, in two independent studies 91,92, researchers utilised genetically modified bacteriophages to fight antibiotic-resistant bacteria by giving them genetic features that specifically target and hinder bacterial defences against antibiotic activity.

In the therapy of cancer, since conventional medicines frequently indiscriminately target tumours and normal tissues, artificially created viruses and organisms that can identify and connect their therapeutic action to pathological signals may be helpful. For example, p53 pathway activity in human cells was put into adenoviruses to control how they replicated.

Engineered Bacteria-Based Platform

Bacteria have long been used in cancer treatment. *Bifidobacterium* and *Clostridium* selectively colonize tumors and reduce their size. Recently synthetic biologists reprogrammed bacteria to sense and respond to a particular cancer state. Most often bacteria are used to deliver a therapeutic molecule directly to the tumor to minimize off-target effects. To target the tumor cells, peptides that can specifically recognize a tumor were expressed on the surfaces of bacteria. Peptides used include an affibody molecule that specifically targets human epidermal growth factor receptor 2 and a synthetic adhesin. The other way is to allow bacteria to sense the tumor microenvironment, for example hypoxia, by building an AND logic gate into bacteria. Then the bacteria only release target therapeutic molecules to the tumor through either lysis or the bacterial secretion system. Lysis has the advantage that it can stimulate the immune system and control growth. Multiple types of secretion systems can be used and other strategies as well. The system is inducible by external signals. Inducers include chemicals, electromagnetic or light waves.

Multiple species and strains are applied in these therapeutics. Most commonly used bacteria are *Salmonella typhimurium*, *Escherichia coli*, *Bifidobacteria*, *Streptococcus*, *Lactobacillus*, *Listeria* and *Bacillus subtilis*. Each of these species have their own property and are unique to cancer therapy in terms of tissue colonization, interaction with immune system and ease of application.

Engineered Yeast-Based Platform

Synthetic biologists are developing genetically modified live yeast that can deliver therapeutic biologic medicines. When orally delivered, these live yeast act like micro-factories and will make therapeutic molecules directly in the gastrointestinal tract. Because yeast are eukaryotic, a key benefit is that they can be administered together with antibiotics. Probiotic yeast expressing human P2Y2 purinergic receptor suppressed intestinal inflammation in mouse models of inflammatory bowel disease. A live S. boulardii yeast delivering a tetra-specific anti-toxin that potently neutralizes Toxin A and Toxin B of *Clostridioides difficile* has been developed. This therapeutic anti-toxin is a fusion of four single-domain antibodies (nanobodies) that potently and broadly neutralize the two major virulence factors of C. difficile at the site of infection in preclinical models. The first in human clinical trial of engineered live yeast for the treatment of Clostridioides difficile infection is anticipated in 2024 and will be sponsored by the developer Fzata, Inc.

Cell-Based Platform

The immune system plays an important role in cancer and can be harnessed to attack cancer cells. Cell-based therapies focus on immunotherapies, mostly by engineering T cells. T cell receptors were engineered and 'trained' to detect cancer epitopes. Chimeric antigen receptors (CARs) are composed of a fragment of an antibody fused to intracellular T cell signaling domains that can activate and trigger proliferation of the cell. Multiple second generation CAR-based therapies have been approved by FDA.

Gene switches were designed to enhance safety of the treatment. Kill switches were developed to terminate the therapy should the patient show severe side effects. Mechanisms can more finely control the system and stop and reactivate it. Since the number of T-cells are important for therapy persistence and severity, growth of T-cells is also controlled to dial the effectiveness and safety of therapeutics.

Although several mechanisms can improve safety and control, limitations include the difficulty of inducing large DNA circuits into the cells and risks associated with introducing foreign components, especially proteins, into cells.

Biofuels, Pharmaceuticals and Biomaterials

The most popular biofuel is ethanol produced from corn or sugar cane, but this method of producing biofuels is troublesome and constrained due to the high agricultural cost and inadequate fuel characteristics of ethanol. An substitute and potential source of renewable energy is microbes that have had their metabolic pathways altered to be more efficient at converting biomass into biofuels. Only if their production costs could be made to match or even exceed those of present fuel production can these techniques be expected to be successful. Related to this, there are several medicines whose pricey manufacturing procedures prevent them from having a larger therapeutic range. The creation of new materials and the microbiological manufacturing of biomaterials would both benefit substantially from novel artificial biology tools.

CRISPR/Cas9

The clustered frequently interspaced short palindromic repetitions (CRISPR)/CRISPR associated (Cas) system is a powerful method of genome engineering in a range of organisms because of its simplicity, modularity, and scalability. In this technique, a guide RNA (gRNA) attracts the CRISPR nuclease Cas9 to a particular spot in the genome, causing a double strand break. Several DNA repair processes, including homology-directed recombination and non-homology end joining, can be used to accomplish the desired genome change (i.e., gene deletion or insertion). Additionally, dCas9 (dead Cas9 or nuclease-deficient Cas9), a Cas9 double mutant (H840A, D10A), has been utilised to control gene expression in bacteria or when linked to a stimulation of suppression site in yeast.

Regulatory Elements

To build and develop biological systems, regulating components including regulators, ribosome-binding sites (RBSs), and terminators are crucial. Despite years of study, there are many various varieties and numbers of promoters and terminators for Escherichia coli, but also for the well-researched model organism Saccharomyces cerevisiae, as well as for other organisms of interest, these tools are quite scarce. Numerous techniques have been invented for the finding and identification of promoters and terminators in order to overcome this constraint, including genome mining, random mutagenesis, hybrid engineering, biophysical modelling, combinatorial design, and rational design.

Organoids

Synthetic biology has been used for organoids, which are lab-grown organs with application to medical research and transplantation.

Bioprinted Organs

3D bioprinting can be used to reconstruct tissue from various regions of the body. The precursor to the adoption of 3D printing in healthcare was a series of trials conducted by researchers at Boston Children's Hospital. The team built replacement urinary bladders by hand for seven patients by constructing scaffolds, then layering the scaffolds with cells from the patients and allowing them to grow. The trials were a success as the patients remained in good health 7 years after implantation, which led a research fellow named Anthony Atala, MD, to search or ways to automate the process. Patients with end-stage bladder disease can now be treated by using bio-engineered bladder tissues to rebuild the damaged organ. This technology can also potentially be applied to bone, skin, cartilage and muscle tissue. Though one long-term goal of 3D bioprinting technology is to reconstruct an entire organ as well as minimize the problem of the lack of organs for transplantation. There has been little success in bioprinting of fully functional organs e.g. liver, skin, meniscus or pancreas. Unlike implantable stents, organs have complex shapes and are significantly harder to bioprint. A bioprinted heart, for example, must not only meet structural requirements, but also vascularization, mechanical load, and electrical signal propagation requirements. In 2022, the first success of a clinical trial for a 3D bioprinted transplant that is made from the patient's own cells, an external ear to treat microtia, was reported.

3D bioprinting contributes to significant advances in the medical field of tissue engineering by allowing for research to be done on innovative materials called biomaterials. Some of the most notable bioengineered substances are usually stronger than the average bodily materials, including soft tissue and bone. These constituents can act as future substitutes, even improvements, for the original body materials. In addition, the Defense Threat Reduction Agency aims to print mini organs such as hearts, livers, and lungs as the potential to test new drugs more accurately and perhaps eliminate the need for testing in animals.

Other Transplants and Induced Regeneration

There is ongoing research and development into synthetic biology based methods for inducing regeneration in humans as well the creation of transplantable artificial organs.

Nanoparticles, Artificial Cells and Micro-Droplets

Synthetic biology can be used for creating nanoparticles which can be used for drug-delivery as well as for other purposes. Complementing research and development seeks to and has created synthetic cells that mimics functions of biological cells. Applications include medicine such as designer-nanoparticles that make blood cells eat away—from the inside out—portions of atherosclerotic plaque that cause heart attacks. Synthetic micro-droplets for algal cells or synergistic algal-bacterial multicellular spheroid microbial reactors, for example, could be used to produce hydrogen as hydrogen economy biotechnology.

Electrogenetics

Mammalian designer cells are engineered by humans to behave a specific way, such as an immune cell that expresses a synthetic receptor designed to combat a specific disease. Electrogenetics is an application of synthetic biology that involves utilizing electrical fields to stimulate a response in engineered cells. Controlling the designer cells can be done with relative ease through the use of common electronic devices, such as smartphones.

Additionally, electrogenetics allows for the possibility of creating devices that are much smaller and compact than devices that use other stimulus through the use of microscopic electrodes. One example of how electrogenetics is used to benefit public health is through stimulating designer cells that are able to produce/deliver therapeutics. This was implemented in $_{Electro}$HEK cells, cells that contain voltage-gated calcium channels that are electrosensitive, meaning that the ion channel can be controlled by electrical conduction between electrodes and the $_{Electro}$HEK cells. The expression levels of the artificial gene that these $_{Electro}$HEK cells contained was shown to be able to be controlled by changing the voltage or electrical pulse length. Further studies have expanded on this robust system, one of which is a beta cell line system designed to control the release of insulin based on electric signals.

Ethics

The creation of new life and the tampering of existing life has raised ethical concerns in the field of synthetic biology and are actively being discussed.

Common ethical questions include:

- Is it morally right to tamper with nature?
- Is one playing God when creating new life?
- What happens if a synthetic organism accidentally escapes?
- What if an individual misuses synthetic biology and creates a harmful entity (e.g., a biological weapon)?

- Who will have control of and access to the products of synthetic biology?
- Who will gain from these innovations? Investors? Medical patients? Industrial farmers?
- Does the patent system allow patents on living organisms? What about parts of organisms, like HIV resistance genes in humans?
- What if a new creation is deserving of moral or legal status?

The ethical aspects of synthetic biology has three main features: biosafety, biosecurity, and the creation of new life forms. Other ethical issues mentioned include the regulation of new creations, patent management of new creations, benefit distribution, and research integrity.

Ethical issues have surfaced for recombinant DNA and genetically modified organism (GMO) technologies and extensive regulations of genetic engineering and pathogen research were in place in many jurisdictions. Amy Gutmann, former head of the Presidential Bioethics Commission, argued that we should avoid the temptation to over-regulate synthetic biology in general, and genetic engineering in particular. According to Gutmann, "Regulatory parsimony is especially important in emerging technologies...where the temptation to stifle innovation on the basis of uncertainty and fear of the unknown is particularly great. The blunt instruments of statutory and regulatory restraint may not only inhibit the distribution of new benefits, but can be counterproductive to security and safety by preventing researchers from developing effective safeguards.".

The "Creation" of Life

One ethical question is whether or not it is acceptable to create new life forms, sometimes known as "playing God". Currently, the creation of new life forms not present in nature is at small-scale, the potential benefits and dangers remain unknown, and careful consideration and oversight are ensured for most studies. Many advocates express the great potential value—to agriculture, medicine, and academic knowledge, among other fields—of creating artificial life forms. Creation of new entities could expand scientific knowledge well beyond what is currently known from studying natural phenomena. Yet there is concern that artificial life forms may reduce nature's "purity" (i.e., nature could be somehow corrupted by human intervention and manipulation) and potentially influence the adoption of more engineering-like principles instead of biodiversity- and nature-focused ideals. Some are also concerned that if an artificial life form were to be released into nature, it could hamper biodiversity by beating out natural species for resources (similar to how algal blooms kill marine species). Another concern involves the ethical treatment of newly created entities if they happen to sense pain, sentience, and self-perception. There is an ongoing debate as to whether such life forms should be granted moral or legal rights, though no consensus exists as to how these rights would be administered or enforced.

Ethical Support for Synthetic Biology

Ethics and moral rationales that support certain applications of synthetic biology include their potential mitigation of substantial global problems of detrimental environmental impacts of conventional agriculture (including meat production), animal welfare, food security, and human health, as well as potential reduction of human labor needs and, via therapies of diseases, reduction of human suffering and prolonged life.

Biosafety and Biocontainment

What is most ethically appropriate when considering biosafety measures? How can accidental introduction of synthetic life in the natural environment be avoided? Much ethical consideration and critical thought has been given to these questions. Biosafety not only refers to biological containment; it also refers to strides taken to protect the public from potentially hazardous biological agents. Even though such concerns are important and remain unanswered, not all products of synthetic biology present concern for biological safety or negative consequences for the environment. It is argued that most synthetic technologies are benign and are incapable of flourishing in the outside world due to their "unnatural" characteristics as there is yet to be an example of a transgenic microbe conferred with a fitness advantage in the wild. In general, existing hazard controls, risk assessment methodologies, and regulations developed for traditional genetically modified organisms (GMOs) are considered to be sufficient for synthetic organisms. "Extrinsic" biocontainment methods in a laboratory context include physical containment through biosafety cabinets and gloveboxes, as well as personal protective equipment. In an agricultural context, they include isolation distances and pollen barriers, similar to methods for biocontainment of GMOs. Synthetic organisms may offer increased hazard control because they can be engineered with "intrinsic" biocontainment methods that limit their growth in an uncontained environment, or prevent horizontal gene transfer to natural organisms.

Examples of intrinsic biocontainment include auxotrophy, biological kill switches, inability of the organism to replicate or to pass modified or synthetic genes to offspring, and the use of xenobiological organisms using alternative biochemistry, for example using artificial xeno nucleic acids (XNA) instead of DNA. Regarding auxotrophy, bacteria and yeast can be engineered to be unable to produce histidine, an important amino acid for all life. Such organisms can thus only be grown on histidine-rich media in laboratory conditions, nullifying fears that they could spread into undesirable areas.

Biosecurity and Bioterrorism

Some ethical issues relate to biosecurity, where biosynthetic technologies could be deliberately used to cause harm to society and/or the environment. Since synthetic biology raises ethical issues and biosecurity issues, humanity must consider and plan on how to deal with potentially harmful creations, and what kinds of ethical measures could possibly be employed to deter nefarious biosynthetic technologies. With the exception of regulating synthetic biology and biotechnology companies, however, the issues are not seen as new because they were raised during the earlier recombinant DNA and genetically modified organism (GMO) debates, and extensive regulations of genetic engineering and pathogen research are already in place in many jurisdictions.

Additionally, the development of synthetic biology tools has made it easier for individuals with less education, training, and access to equipment to modify and use pathogenic organisms as bioweapons. This increases the threat of bioterrorism, especially as terrorist groups become aware of the significant social, economic, and political disruption caused by pandemics like COVID-19. As new techniques are developed in the field of synthetic biology, the risk of bioterrorism is likely to continue to grow.

Juan Zarate, who served as Deputy National Security Advisor for Combating Terrorism from 2005 to 2009, noted that "the severity and extreme disruption of a novel coronavirus will likely spur the imagination of the most creative and dangerous groups and individuals to reconsider bioterrorist attacks."

European Union

The European Union-funded project SYNBIOSAFE has issued reports on how to manage synthetic biology. A 2007 paper identified key issues in safety, security, ethics, and the science-society interface, which the project defined as public education and ongoing dialogue among scientists, businesses, government and ethicists. The key security issues that SYNBIOSAFE identified involved engaging companies that sell synthetic DNA and the biohacking community of amateur biologists. Key ethical issues concerned the creation of new life forms.

A subsequent report focused on biosecurity, especially the so-called dual-use challenge. For example, while synthetic biology may lead to more efficient production of medical treatments, it may also lead to synthesis or modification of harmful pathogens (e.g., smallpox). The biohacking community remains a source of special concern, as the distributed and diffuse nature of open-source biotechnology makes it difficult to track, regulate or mitigate potential concerns over biosafety and biosecurity. COSY, another European initiative, focuses on public perception and communication. To better communicate synthetic biology and its societal ramifications to a broader public, COSY and SYNBIOSAFE published *SYNBIOSAFE*, a 38-minute documentary film, in October 2009.

The International Association Synthetic Biology has proposed self-regulation. This proposes specific measures that the synthetic biology industry, especially DNA synthesis companies, should implement. In 2007, a group led by scientists from leading DNA-synthesis companies published a "practical plan for developing an effective oversight framework for the DNA-synthesis industry".

United States

In January 2009, the Alfred P. Sloan Foundation funded the Woodrow Wilson Center, the Hastings Center, and the J. Craig Venter Institute to examine the public perception, ethics and policy implications of synthetic biology.

On July 9–10, 2009, the National Academies' Committee of Science, Technology & Law convened a symposium on "Opportunities and Challenges in the Emerging Field of Synthetic Biology". After the publication of the first synthetic genome and the accompanying media coverage about "life" being created, President Barack Obama established the Presidential Commission for the Study of Bioethical Issues to study synthetic biology. The commission convened a series of meetings, and issued a report in December 2010 titled "New Directions: The Ethics of Synthetic Biology and Emerging Technologies." The commission stated that "while Venter's achievement marked a significant technical advance in demonstrating that a relatively large genome could be accurately synthesized and substituted for another, it did not amount to the "creation of life". It noted that synthetic biology is an emerging field, which creates potential risks and rewards. The commission did not recommend policy or oversight changes and called for continued funding of the research and new funding for monitoring, study of emerging ethical issues and public education.

Synthetic biology, as a major tool for biological advances, results in the "potential for developing biological weapons, possible unforeseen negative impacts on human health ... and any potential environmental impact". The proliferation of such technology could also make the production of biological and chemical weapons available to a wider array of state and non-state actors. These security issues may be avoided by regulating industry uses of biotechnology through policy legislation. Federal guidelines on genetic manipulation are being proposed by "the President's Bioethics Commission ... in response to the announced creation of a self-replicating cell from a chemically synthesized genome, put forward 18 recommendations not only for regulating the science ... for educating the public".

Opposition

On March 13, 2012, over 100 environmental and civil society groups, including

Friends of the Earth, the International Center for Technology Assessment, and the ETC Group, issued the manifesto *The Principles for the Oversight of Synthetic Biology*. This manifesto calls for a worldwide moratorium on the release and commercial use of synthetic organisms until more robust regulations and rigorous biosafety measures are established.

The groups specifically call for an outright ban on the use of synthetic biology on the human genome or human microbiome. Richard Lewontin wrote that some of the safety tenets for oversight discussed in *The Principles for the Oversight of Synthetic Biology* are reasonable, but that the main problem with the recommendations in the manifesto is that "the public at large lacks the ability to enforce any meaningful realization of those recommendations".

Health and Safety

The hazards of synthetic biology include biosafety hazards to workers and the public, biosecurity hazards stemming from deliberate engineering of organisms to cause harm, and environmental hazards. The biosafety hazards are similar to those for existing fields of biotechnology, mainly exposure to pathogens and toxic chemicals, although novel synthetic organisms may have novel risks. For biosecurity, there is concern that synthetic or redesigned organisms could theoretically be used for bioterrorism. Potential risks include recreating known pathogens from scratch, engineering existing pathogens to be more dangerous, and engineering microbes to produce harmful biochemicals. Lastly, environmental hazards include adverse effects on biodiversity and ecosystem services, including potential changes to land use resulting from agricultural use of synthetic organisms. Synthetic biology is an example of a dual-use technology with the potential to be used in ways that could intentionally or unintentionally harm humans and/or damage the environment. Often "scientists, their host institutions and funding bodies" consider whether the planned research could be misused and sometimes implement measures to reduce the likelihood of misuse.

Existing risk analysis systems for GMOs are generally considered sufficient for synthetic organisms, although there may be difficulties for an organism built "bottom-up" from individual genetic sequences. Synthetic biology generally falls under existing regulations for GMOs and biotechnology in general, and any regulations that exist for downstream commercial products, although there are generally no regulations in any jurisdiction that are specific to synthetic biology.

7.5 MEMBRANE ENGINEERING FOR IMPROVED EFFLUX

The structure and composition of the cell membrane play a crucial role in metabolite efflux. Advances in membrane engineering have enabled scientists to modify cell membrane properties to enhance transport efficiency. Membrane fluidity, thickness, and composition can all influence the efficiency of metabolite export, particularly for hydrophobic compounds like fatty acids and biofuels.

Membrane engineering approaches often focus on optimizing the lipid composition of the cell membrane to facilitate metabolite diffusion. For instance, increasing the unsaturated fatty acid content in the membrane can enhance membrane fluidity, enabling faster diffusion of hydrophobic molecules. This approach has been used in *Saccharomyces cerevisiae* to improve ethanol tolerance and efflux, enhancing the strain's suitability for industrial-scale bioethanol production.

Additionally, proteins like porins—channels that facilitate the passive transport of molecules—have been modified to increase their permeability to specific compounds. Engineering porins or other membrane channels to increase efflux efficiency is an emerging strategy that complements traditional transporter-based methods, especially for large or hydrophobic molecules.

7.6 FUTURE PROSPECTS: INTEGRATED EFFLUX CONTROL IN MICROBIAL CELL FACTORIES

The future of metabolite efflux control in microbial cell factories is likely to involve highly integrated systems that combine genetic, synthetic biology, and membrane engineering approaches. By integrating multiple strategies, researchers can develop comprehensive solutions that allow for fine-tuned control over efflux, tailored to the specific needs of each microbial production system.

In the future, machine learning and AI are expected to play a significant role in optimizing efflux systems. Machine learning algorithms can analyze large datasets from genetic, transcriptomic, and metabolic studies to identify the best candidates for efflux control and predict the impacts of various modifications. These insights will streamline the design of customized efflux pathways, allowing for rapid optimization of microbial strains.

Another promising area is the development of modular efflux systems, where transporter proteins, membrane modifications, and biosensors are designed as interchangeable modules. Such modularity would enable researchers to quickly adapt microbial cell factories to new products by swapping in the appropriate efflux modules, reducing the time required to bring new products to market.

7.7 CONCLUSION

The control of metabolite efflux is a cornerstone of modern microbial cell factory design, ensuring efficient and sustainable production of a wide range of bioproducts. Advances in genetic engineering, synthetic biology, and membrane engineering have made it possible to tailor efflux systems to meet the specific demands of different industrial applications. With the integration of smart biosensors and computational models, the precision of efflux control continues to improve, reducing metabolic burden and enhancing productivity.

Looking ahead, the future of efflux control in microbial cell factories is bright, driven by the integration of emerging technologies and multi-disciplinary approaches. As AI and machine learning become more prevalent, the potential for rapid optimization and customization of efflux systems will further expand the applications of microbial cell factories, offering a sustainable and efficient pathway to meet the growing demand for bio-based products.

Through these advances, microbial cell factories will play an increasingly pivotal role in industries ranging from healthcare to energy, supporting a sustainable and bio-based economy.

SECTION III:

FERMENTATION MICROBIOLOGY AND BIOTECHNOLOGY: METABOLIC, ENZYMATIC, AND GENETIC ENGINEERING OF MICROORGANISMS

CHAPTER 8

INCREASING TOLERANCE AND ROBUSTNESS OF INDUSTRIAL MICROORGANISMS: CURRENT STRATEGIES AND FUTURE PROSPECTS

In industrial biotechnology, microorganisms such as bacteria, yeast, and fungi are harnessed for the production of biofuels, pharmaceuticals, chemicals, and food ingredients. However, the conditions within large-scale bioreactors can be stressful for microbial cells, often limiting productivity and yield. Stressors such as high product concentrations, toxic intermediates, pH fluctuations, and high temperatures can inhibit growth, reduce metabolic efficiency, and ultimately impact the stability of production processes. Improving the tolerance and robustness of industrial microorganisms is crucial for efficient bioprocesses and for maintaining productivity under adverse conditions.

8.1 IMPORTANCE OF TOLERANCE AND ROBUSTNESS IN INDUSTRIAL MICROORGANISMS

For industrial microorganisms, robustness refers to their ability to maintain stable productivity under varying environmental conditions, while tolerance is their ability to withstand specific stress factors such as high concentrations of a target product. Microbial strains with high tolerance and robustness are essential for consistent production in industrial bioreactors, which can expose cells to harsh conditions that are much different from those in natural habitats or laboratory settings. For instance, microbial production of ethanol requires tolerance to high alcohol concentrations, which can be toxic to cells. Similarly, production of organic acids like lactic acid or acetic acid can lead to a low pH environment, which can disrupt cellular functions.

Enhancing microbial tolerance to these conditions not only improves process efficiency but also reduces costs associated with bioprocess optimization and downstream processing. Robust strains can often achieve higher yields, have longer lifespans in bioreactors, and require fewer inputs to maintain stability, making the entire production process more sustainable.

8.2 GENETIC ENGINEERING APPROACHES FOR ENHANCED TOLERANCE

8.2.1 Genetic Engineering: An Overview

Genetic engineering, also called genetic modification or genetic manipulation, is the modification and manipulation of an organism's genes using technology. It is a set of technologies used to change the genetic makeup of cells, including the transfer of genes within and across species boundaries to produce improved or novel organisms.

New DNA is obtained by either isolating and copying the genetic material of interest using recombinant DNA methods or by artificially synthesising the DNA. A construct is usually created and used to insert this DNA into the host organism. The first recombinant DNA molecule was made by Paul Berg in 1972 by combining DNA from the monkey virus SV40 with the lambda virus.

As well as inserting genes, the process can be used to remove, or "knock out", genes. The new DNA can be inserted randomly, or targeted to a specific part of the genome.

An organism that is generated through genetic engineering is considered to be genetically modified (GM) and the resulting entity is a genetically modified organism (GMO). The first GMO was a bacterium generated by Herbert Boyer and Stanley Cohen in 1973. Rudolf Jaenisch created the first GM animal when he inserted foreign DNA into a mouse in 1974. The first company to focus on genetic engineering, Genentech, was founded in 1976 and started the production of human proteins. Genetically engineered human insulin was produced in 1978 and insulin-producing bacteria were commercialised in 1982. Genetically modified food has been sold since 1994, with the release of the Flavr Savr tomato. The Flavr Savr was engineered to have a longer shelf life, but most current GM crops are modified to increase resistance to insects and herbicides. GloFish, the first GMO designed as a pet, was sold in the United States in December 2003. In 2016 salmon modified with a growth hormone were sold.

Genetic engineering has been applied in numerous fields including research, medicine, industrial biotechnology and agriculture. In research, GMOs are used to study gene function and expression through loss of function, gain of function, tracking and expression experiments. By knocking out genes responsible for certain conditions it is possible to create animal model organisms of human diseases. As well as producing hormones, vaccines and other drugs, genetic engineering has the potential to cure genetic diseases through gene therapy. Chinese hamster ovary (CHO) cells are used in industrial genetic engineering. Additionally mRNA vaccines are made through genetic engineering to treat viruses such as COVID-19. The same techniques that are used to produce drugs can also have industrial applications such as producing enzymes for laundry detergent, cheeses and other products.

The rise of commercialised genetically modified crops has provided economic benefit to farmers in many different countries, but has also been the source of most of the controversy surrounding the technology. This has been present since its early use; the first field trials were destroyed by anti-GM activists. Although there is a scientific consensus that currently available food derived from GM crops poses no greater risk to human health than conventional food, critics consider GM food safety a leading concern. Gene flow, impact on non-target organisms, control of the food supply and intellectual property rights have also been raised as potential issues. These concerns have led to the development of a regulatory framework, which started in 1975. It has led to an international treaty, the Cartagena Protocol on Biosafety, that was adopted in 2000. Individual countries have developed their own regulatory systems regarding GMOs, with the most marked differences occurring between the United States and Europe.

Genetic engineering is a process that alters the genetic structure of an organism by either removing or introducing DNA, or modifying existing genetic material in situ. Unlike traditional animal and plant breeding, which involves doing multiple crosses and then selecting for the organism with the desired phenotype, genetic engineering takes the gene directly from one organism and delivers it to the other. This is much faster, can be used to insert any genes from any organism (even ones from different domains) and prevents other undesirable genes from also being added.

Genetic engineering could potentially fix severe genetic disorders in humans by replacing the defective gene with a functioning one. It is an important tool in research that allows the function of specific genes to be studied. Drugs, vaccines and other products have been harvested from organisms engineered to produce them. Crops have been developed that aid food security by increasing yield, nutritional value and tolerance to environmental stresses.

The DNA can be introduced directly into the host organism or into a cell that is then fused or hybridised with the host. This relies on recombinant nucleic acid techniques to form new combinations of heritable genetic material followed by the incorporation of that material either indirectly through a vector system or directly through micro-injection, macro-injection or micro-encapsulation.

Genetic engineering does not normally include traditional breeding, in vitro fertilisation, induction of polyploidy, mutagenesis and cell fusion techniques that do not use recombinant nucleic acids or a genetically modified organism in the process. However, some broad definitions of genetic engineering include selective breeding. Cloning and stem cell research, although not considered genetic engineering, are closely related and genetic engineering can be used within them. Synthetic biology is an emerging discipline that takes genetic engineering a step further by introducing artificially synthesised material into an organism.

Plants, animals or microorganisms that have been changed through genetic engineering are termed genetically modified organisms or GMOs. If genetic material from another species is added to the host, the resulting organism is called transgenic.

If genetic material from the same species or a species that can naturally breed with the host is used the resulting organism is called cisgenic. If genetic engineering is used to remove genetic material from the target organism the resulting organism is termed a knockout organism. In Europe genetic modification is synonymous with genetic engineering while within the United States of America and Canada genetic modification can also be used to refer to more conventional breeding methods.

History

Humans have altered the genomes of species for thousands of years through selective breeding, or artificial selection as contrasted with natural selection. More recently, mutation breeding has used exposure to chemicals or radiation to produce a high frequency of random mutations, for selective breeding purposes. Genetic engineering as the direct manipulation of DNA by humans outside breeding and mutations has only existed since the 1970s.

The term "genetic engineering" was coined by the Russian-born geneticist Nikolay Timofeev-Ressovsky in his 1934 paper "The Experimental Production of Mutations", published in the British journal Biological Reviews. Jack Williamson used the term in his science fiction novel Dragon's Island, published in 1951 – one year before DNA's role in heredity was confirmed by Alfred Hershey and Martha Chase, and two years before James Watson and Francis Crick showed that the DNA molecule has a double-helix structure – though the general concept of direct genetic manipulation was explored in rudimentary form in Stanley G. Weinbaum's 1936 science fiction story *Proteus Island.*

In 1972, Paul Berg created the first recombinant DNA molecules by combining DNA from the monkey virus SV40 with that of the lambda virus. In 1973 Herbert Boyer and Stanley Cohen created the first transgenic organism by inserting antibiotic resistance genes into the plasmid of an *Escherichia coli* bacterium. A year later Rudolf Jaenisch created a transgenic mouse by introducing foreign DNA into its embryo, making it the world's first transgenic animal These achievements led to concerns in the scientific community about potential risks from genetic engineering, which were first discussed in depth at the Asilomar Conference in 1975. One of the main recommendations from this meeting was that government oversight of recombinant DNA research should be established until the technology was deemed safe.

In 1976 Genentech, the first genetic engineering company, was founded by Herbert Boyer and Robert Swanson and a year later the company produced a human protein (somatostatin) in *E. coli*. Genentech announced the production of genetically engineered human insulin in 1978. In 1980, the U.S. Supreme Court in the *Diamond v. Chakrabarty* case ruled that genetically altered life could be patented. The insulin produced by bacteria was approved for release by the Food and Drug Administration (FDA) in 1982.

In 1983, a biotech company, Advanced Genetic Sciences (AGS) applied for U.S. government authorisation to perform field tests with the ice-minus strain of *Pseudomonas syringae* to protect crops from frost, but environmental groups and protestors delayed the field tests for four years with legal challenges. In 1987, the ice-minus strain of *P. syringae* became the first genetically modified organism (GMO) to be released into the environment when a strawberry field and a potato field in California were sprayed with it. Both test fields were attacked by activist groups the night before the tests occurred: "The world's first trial site attracted the world's first field trasher".

The first field trials of genetically engineered plants occurred in France and the US in 1986, tobacco plants were engineered to be resistant to herbicides. The People's Republic of China was the first country to commercialise transgenic plants, introducing a virus-resistant tobacco in 1992. In 1994 Calgene attained approval to commercially release the first genetically modified food, the Flavr Savr, a tomato engineered to have a longer shelf life. In 1994, the European Union approved tobacco engineered to be resistant to the herbicide bromoxynil, making it the first genetically engineered crop commercialised in Europe. In 1995, Bt potato was approved safe by the Environmental Protection Agency, after having been approved by the FDA, making it the first pesticide producing crop to be approved in the US. In 2009 11 transgenic crops were grown commercially in 25 countries, the largest of which by area grown were the US, Brazil, Argentina, India, Canada, China, Paraguay and South Africa.

In 2010, scientists at the J. Craig Venter Institute created the first synthetic genome and inserted it into an empty bacterial cell. The resulting bacterium, named Mycoplasma laboratorium, could replicate and produce proteins. Four years later this was taken a step further when a bacterium was developed that replicated a plasmid containing a unique base pair, creating the first organism engineered to use an expanded genetic alphabet. In 2012, Jennifer Doudna and Emmanuelle Charpentier collaborated to develop the CRISPR/Cas9 system, a technique which can be used to easily and specifically alter the genome of almost any organism.

Process

Creating a GMO is a multi-step process. Genetic engineers must first choose what gene they wish to insert into the organism. This is driven by what the aim is for the resultant organism and is built on earlier research. Genetic screens can be carried out to determine potential genes and further tests then used to identify the best candidates. The development of microarrays, transcriptomics and genome sequencing has made it much easier to find suitable genes. Luck also plays its part; the Roundup Ready gene was discovered after scientists noticed a bacterium thriving in the presence of the herbicide.

Gene Isolation and Cloning

The next step is to isolate the candidate gene. The cell containing the gene is opened and the DNA is purified. The gene is separated by using restriction enzymes to cut the DNA into fragments or polymerase chain reaction (PCR) to amplify up the gene segment. These segments can then be extracted through gel electrophoresis. If the chosen gene or the donor organism's genome has been well studied it may already be accessible from a genetic library. If the DNA sequence is known, but no copies of the gene are available, it can also be artificially synthesised. Once isolated the gene is ligated into a plasmid that is then inserted into a bacterium. The plasmid is replicated when the bacteria divide, ensuring unlimited copies of the gene are available. The RK2 plasmid is notable for its ability to replicate in a wide variety of single-celled organisms, which makes it suitable as a genetic engineering tool.

Before the gene is inserted into the target organism it must be combined with other genetic elements. These include a promoter and terminator region, which initiate and end transcription. A selectable marker gene is added, which in most cases confers antibiotic resistance, so researchers can easily determine which cells have been successfully transformed. The gene can also be modified at this stage for better expression or effectiveness. These manipulations are carried out using recombinant DNA techniques, such as restriction digests, ligations and molecular cloning.

Inserting DNA into the Host Genome

There are a number of techniques used to insert genetic material into the host genome. Some bacteria can naturally take up foreign DNA. This ability can be induced in other bacteria via stress (e.g. thermal or electric shock), which increases the cell membrane's permeability to DNA; up-taken DNA can either integrate with the genome or exist as extrachromosomal DNA.

DNA is generally inserted into animal cells using microinjection, where it can be injected through the cell's nuclear envelope directly into the nucleus, or through the use of viral vectors.

Plant genomes can be engineered by physical methods or by use of *Agrobacterium* for the delivery of sequences hosted in T-DNA binary vectors. In plants the DNA is often inserted using *Agrobacterium*-mediated transformation, taking advantage of the *Agrobacteriums* T-DNA sequence that allows natural insertion of genetic material into plant cells. Other methods include biolistics, where particles of gold or tungsten are coated with DNA and then shot into young plant cells, and electroporation, which involves using an electric shock to make the cell membrane permeable to plasmid DNA.

As only a single cell is transformed with genetic material, the organism must be regenerated from that single cell. In plants this is accomplished through the use of tissue culture. In animals it is necessary to ensure that the inserted DNA is present in the embryonic stem cells. Bacteria consist of a single cell and reproduce clonally so regeneration is not necessary. Selectable markers are used to easily differentiate transformed from untransformed cells. These markers are usually present in the transgenic organism, although a number of strategies have been developed that can remove the selectable marker from the mature transgenic plant.

Further testing using PCR, Southern hybridization, and DNA sequencing is conducted to confirm that an organism contains the new gene. These tests can also confirm the chromosomal location and copy number of the inserted gene. The presence of the gene does not guarantee it will be expressed at appropriate levels in the target tissue so methods that look for and measure the gene products (RNA and protein) are also used. These include northern hybridisation, quantitative RT-PCR, Western blot, immunofluorescence, ELISA and phenotypic analysis.

The new genetic material can be inserted randomly within the host genome or targeted to a specific location. The technique of gene targeting uses homologous recombination to make desired changes to a specific endogenous gene. This tends to occur at a relatively low frequency in plants and animals and generally requires the use of selectable markers. The frequency of gene targeting can be greatly enhanced through genome editing. Genome editing uses artificially engineered nucleases that create specific double-stranded breaks at desired locations in the genome, and use the cell's endogenous mechanisms to repair the induced break by the natural processes of homologous recombination and nonhomologous end-joining. There are four families of engineered nucleases: meganucleases, zinc finger nucleases, transcription activator-like effector nucleases (TALENs), and the Cas9-guideRNA system (adapted from CRISPR). TALEN and CRISPR are the two most commonly used and each has its own advantages. TALENs have greater target specificity, while CRISPR is easier to design and more efficient. In addition to enhancing gene targeting, engineered nucleases can be used to introduce mutations at endogenous genes that generate a gene knockout.

Applications

Genetic engineering has applications in medicine, research, industry and agriculture and can be used on a wide range of plants, animals and microorganisms. Bacteria, the first organisms to be genetically modified, can have plasmid DNA inserted containing new genes that code for medicines or enzymes that process food and other substrates. Plants have been modified for insect protection, herbicide resistance, virus resistance, enhanced nutrition, tolerance to environmental pressures and the production of edible vaccines. Most commercialised GMOs are insect resistant or herbicide tolerant crop plants. Genetically modified animals have been used for research, model animals and the production of agricultural or pharmaceutical products. The genetically modified animals include animals with genes knocked out, increased susceptibility to disease, hormones for extra growth and the ability to express proteins in their milk.

Medicine

Genetic engineering has many applications to medicine that include the manufacturing of drugs, creation of model animals that mimic human conditions and gene therapy. One of the earliest uses of genetic engineering was to mass-produce human insulin in bacteria. This application has now been applied to human growth hormones, follicle stimulating hormones (for treating infertility), human albumin, monoclonal antibodies, antihemophilic factors, vaccines and many other drugs. Mouse hybridomas, cells fused together to create monoclonal antibodies, have been adapted through genetic engineering to create human monoclonal antibodies. Genetically engineered viruses are being developed that can still confer immunity, but lack the infectious sequences.

Genetic engineering is also used to create animal models of human diseases. Genetically modified mice are the most common genetically engineered animal model. They have been used to study and model cancer (the oncomouse), obesity, heart disease, diabetes, arthritis, substance abuse, anxiety, aging and Parkinson disease. Potential cures can be tested against these mouse models.

Gene therapy is the genetic engineering of humans, generally by replacing defective genes with effective ones. Clinical research using somatic gene therapy has been conducted with several diseases, including X-linked SCID, chronic lymphocytic leukemia (CLL), and Parkinson's disease.

In 2012, Alipogene tiparvovec became the first gene therapy treatment to be approved for clinical use. In 2015 a virus was used to insert a healthy gene into the skin cells of a boy suffering from a rare skin disease, epidermolysis bullosa, in order to grow, and then graft healthy skin onto 80 percent of the boy's body which was affected by the illness.

Germline gene therapy would result in any change being inheritable, which has raised concerns within the scientific community. In 2015, CRISPR was used to edit the DNA of non-viable human embryos, leading scientists of major world academies to call for a moratorium on inheritable human genome edits. There are also concerns that the technology could be used not just for treatment, but for enhancement, modification or alteration of a human beings' appearance, adaptability, intelligence, character or behavior. The distinction between cure and enhancement can also be difficult to establish. In November 2018, He Jiankui announced that he had edited the genomes of two human embryos, to attempt to disable the *CCR5* gene, which codes for a receptor that HIV uses to enter cells. The work was widely condemned as unethical, dangerous, and premature. Currently, germline modification is banned in 40 countries. Scientists that do this type of research will often let embryos grow for a few days without allowing it to develop into a baby.

Researchers are altering the genome of pigs to induce the growth of human organs, with the aim of increasing the success of pig to human organ transplantation. Scientists are creating "gene drives", changing the genomes of mosquitoes to make them immune to malaria, and then looking to spread the genetically altered mosquitoes throughout the mosquito population in the hopes of eliminating the disease.

Research

Genetic engineering is an important tool for natural scientists, with the creation of transgenic organisms one of the most important tools for analysis of gene function. Genes and other genetic information from a wide range of organisms can be inserted into bacteria for storage and modification, creating genetically modified bacteria in the process. Bacteria are cheap, easy to grow, clonal, multiply quickly, relatively easy to transform and can be stored at -80 °C almost indefinitely. Once a gene is isolated it can be stored inside the bacteria providing an unlimited supply for research.

Organisms are genetically engineered to discover the functions of certain genes. This could be the effect on the phenotype of the organism, where the gene is expressed or what other genes it interacts with. These experiments generally involve loss of function, gain of function, tracking and expression.

- Loss of function experiments, such as in a gene knockout experiment, in which an organism is engineered to lack the activity of one or more genes. In a simple knockout a copy of the desired gene has been altered to make it non-functional. Embryonic stem cells incorporate the altered gene, which replaces the already present functional copy. These stem cells are injected into blastocysts, which are implanted into surrogate mothers. This allows the experimenter to analyse the defects caused by this mutation and thereby determine the role of particular genes. It is used especially frequently in developmental biology. When this is done by creating a library of genes with point mutations at every position in the area of interest, or even every position in the whole gene, this is called "scanning mutagenesis". The simplest method, and the first to be used, is "alanine scanning", where every position in turn is mutated to the unreactive amino acid alanine.

- Gain of function experiments, the logical counterpart of knockouts. These are sometimes performed in conjunction with knockout experiments to more finely establish the function of the desired gene. The process is much the same as that in knockout engineering, except that the construct is designed to increase the function of the gene, usually by providing extra copies of the gene or inducing synthesis of the protein more frequently. Gain of function is used to tell whether or not a protein is sufficient for a function, but does not always mean it is required, especially when dealing with genetic or functional redundancy.
- Tracking experiments, which seek to gain information about the localisation and interaction of the desired protein. One way to do this is to replace the wild-type gene with a 'fusion' gene, which is a juxtaposition of the wild-type gene with a reporting element such as green fluorescent protein (GFP) that will allow easy visualisation of the products of the genetic modification. While this is a useful technique, the manipulation can destroy the function of the gene, creating secondary effects and possibly calling into question the results of the experiment. More sophisticated techniques are now in development that can track protein products without mitigating their function, such as the addition of small sequences that will serve as binding motifs to monoclonal antibodies.
- Expression studies aim to discover where and when specific proteins are produced. In these experiments, the DNA sequence before the DNA that codes for a protein, known as a gene's promoter, is reintroduced into an organism with the protein coding region replaced by a reporter gene such as GFP or an enzyme that catalyses the production of a dye. Thus the time and place where a particular protein is produced can be observed. Expression studies can be taken a step further by altering the promoter to find which pieces are crucial for the proper expression of the gene and are actually bound by transcription factor proteins; this process is known as promoter bashing.

Industrial

Organisms can have their cells transformed with a gene coding for a useful protein, such as an enzyme, so that they will overexpress the desired protein. Mass quantities of the protein can then be manufactured by growing the transformed organism in bioreactor equipment using industrial fermentation, and then purifying the protein. Some genes do not work well in bacteria, so yeast, insect cells or mammalian cells can also be used. These techniques are used to produce medicines such as insulin, human growth hormone, and vaccines, supplements such as tryptophan, aid in the production of food (chymosin in cheese making) and fuels. Other applications with genetically engineered bacteria could involve making them perform tasks outside their natural cycle, such as making biofuels, cleaning up oil spills, carbon and other toxic waste and detecting arsenic in drinking water. Certain genetically modified microbes can also be used in biomining and bioremediation, due to their ability to extract heavy metals from their environment and incorporate them into compounds that are more easily recoverable.

In materials science, a genetically modified virus has been used in a research laboratory as a scaffold for assembling a more environmentally friendly lithium-ion battery. Bacteria have also been engineered to function as sensors by expressing a fluorescent protein under certain environmental conditions.

Agriculture

One of the best-known and controversial applications of genetic engineering is the creation and use of genetically modified crops or genetically modified livestock to produce genetically modified food. Crops have been developed to increase production, increase tolerance to abiotic stresses, alter the composition of the food, or to produce novel products.

The first crops to be released commercially on a large scale provided protection from insect pests or tolerance to herbicides. Fungal and virus resistant crops have also been developed or are in development. This makes the insect and weed management of crops easier and can indirectly increase crop yield. GM crops that directly improve yield by accelerating growth or making the plant more hardy (by improving salt, cold or drought tolerance) are also under development. In 2016 Salmon have been genetically modified with growth hormones to reach normal adult size much faster.

GMOs have been developed that modify the quality of produce by increasing the nutritional value or providing more industrially useful qualities or quantities. The Amflora potato produces a more industrially useful blend of starches. Soybeans and canola have been genetically modified to produce more healthy oils. The first commercialised GM food was a tomato that had delayed ripening, increasing its shelf life.

Plants and animals have been engineered to produce materials they do not normally make. Pharming uses crops and animals as bioreactors to produce vaccines, drug intermediates, or the drugs themselves; the useful product is purified from the harvest and then used in the standard pharmaceutical production process. Cows and goats have been engineered to express drugs and other proteins in their milk, and in 2009 the FDA approved a drug produced in goat milk.

Other Applications

Genetic engineering has potential applications in conservation and natural area management. Gene transfer through viral vectors has been proposed as a means of controlling invasive species as well as vaccinating threatened fauna from disease. Transgenic trees have been suggested as a way to confer resistance to pathogens in wild populations. With the increasing risks of maladaptation in organisms as a result of climate change and other perturbations, facilitated adaptation through gene tweaking could be one solution to reducing extinction risks. Applications of genetic engineering in conservation are thus far mostly theoretical and have yet to be put into practice.

Genetic engineering is also being used to create microbial art. Some bacteria have been genetically engineered to create black and white photographs. Novelty items such as lavender-colored carnations, blue roses, and glowing fish, have also been produced through genetic engineering.

Regulation

The regulation of genetic engineering concerns the approaches taken by governments to assess and manage the risks associated with the development and release of GMOs. The development of a regulatory framework began in 1975, at Asilomar, California. The Asilomar meeting recommended a set of voluntary guidelines regarding the use of recombinant technology. As the technology improved the US established a committee at the Office of Science and Technology, which assigned regulatory approval of GM food to the USDA, FDA and EPA. The Cartagena Protocol on Biosafety, an international treaty that governs the transfer, handling, and use of GMOs, was adopted on 29 January 2000. One hundred and fifty-seven countries are members of the Protocol, and many use it as a reference point for their own regulations.

The legal and regulatory status of GM foods varies by country, with some nations banning or restricting them, and others permitting them with widely differing degrees of regulation. Some countries allow the import of GM food with authorisation, but either do not allow its cultivation (Russia, Norway, Israel) or have provisions for cultivation even though no GM products are yet produced (Japan, South Korea). Most countries that do not allow GMO cultivation do permit research. Some of the most marked differences occur between the US and Europe. The US policy focuses on the product (not the process), only looks at verifiable scientific risks and uses the concept of substantial equivalence. The European Union by contrast has possibly the most stringent GMO regulations in the world. All GMOs, along with irradiated food, are considered "new food" and subject to extensive, case-by-case, science-based food evaluation by the European Food Safety Authority. The criteria for authorisation fall in four broad categories: "safety", "freedom of choice", "labelling", and "traceability". The level of regulation in other countries that cultivate GMOs lie in between Europe and the United States.

Table 1: Regulatory Agencies by Geographical Region

Region	Regulators	Notes
US	USDA, FDA and EPA	
Europe	European Food Safety Authority	
Canada	Health Canada and the Canadian Food Inspection Agency	Regulated products with novel features regardless of method of origin
Africa	Common Market for Eastern and Southern Africa	Final decision lies with each individual country.
China	Office of Agricultural Genetic Engineering Biosafety Administration	
India	Institutional Biosafety Committee, Review Committee on Genetic Manipulation and Genetic Engineering Approval Committee	
Argentina	National Agricultural Biotechnology Advisory Committee (environmental impact), the National Service of Health and Agrifood Quality (food safety) and the National Agribusiness Direction (effect on trade)	Final decision made by the Secretariat of Agriculture, Livestock, Fishery and Food.
Brazil	National Biosafety Technical Commission (environmental and food safety) and the Council of Ministers (commercial and economical issues)	
Australia	Office of the Gene Technology Regulator (oversees all GM products), Therapeutic Goods Administration (GM medicines) and Food Standards Australia New Zealand (GM food).	The individual state governments can then assess the impact of release on markets and trade and apply further legislation to control approved genetically modified products.

One of the key issues concerning regulators is whether GM products should be labeled. The European Commission says that mandatory labeling and traceability are needed to allow for informed choice, avoid potential false advertising and facilitate the withdrawal of products if adverse effects on health or the environment are discovered. The American Medical Association and the American Association for the Advancement of Science say that absent scientific evidence of harm even voluntary labeling is misleading and will falsely alarm consumers. Labeling of GMO products in the marketplace is required in 64 countries. Labeling can be mandatory up to a threshold GM content level (which varies between countries) or voluntary. In Canada and the US labeling of GM food is voluntary, while in Europe all food (including processed food) or feed which contains greater than 0.9% of approved GMOs must be labelled.

Genetically Modified Food Controversies

Critics have objected to the use of genetic engineering on several grounds, including ethical, ecological and economic concerns. Many of these concerns involve GM crops and whether food produced from them is safe and what impact growing them will have on the environment. These controversies have led to litigation, international trade disputes, and protests, and to restrictive regulation of commercial products in some countries. Accusations that scientists are "playing God" and other religious issues have been ascribed to the technology from the beginning. Other ethical issues raised include the patenting of life, the use of intellectual property rights, the level of labeling on products, control of the food supply and the objectivity of the regulatory process. Although doubts have been raised, economically most studies have found growing GM crops to be beneficial to farmers.

Gene flow between GM crops and compatible plants, along with increased use of selective herbicides, can increase the risk of "superweeds" developing. Other environmental concerns involve potential impacts on non-target organisms, including soil microbes, and an increase in secondary and resistant insect pests. Many of the environmental impacts regarding GM crops may take many years to be understood and are also evident in conventional agriculture practices. With the commercialisation of genetically modified fish there are concerns over what the environmental consequences will be if they escape.

There are three main concerns over the safety of genetically modified food: whether they may provoke an allergic reaction; whether the genes could transfer from the food into human cells; and whether the genes not approved for human consumption could outcross to other crops. There is a scientific consensus that currently available food derived from GM crops poses no greater risk to human health than conventional food, but that each GM food needs to be tested on a case-by-case basis before introduction. Nonetheless, members of the public are less likely than scientists to perceive GM foods as safe.

8.2.2 Genetic Engineering Strategies to Improve Tolerance

Genetic engineering is one of the primary approaches for improving microbial tolerance. By altering specific genes, scientists can improve a microorganism's ability to withstand stressors such as solvents, acids, osmotic pressure, and temperature. This approach often involves either overexpressing genes that confer stress resistance or knocking out genes that contribute to sensitivity.

For instance, to increase ethanol tolerance in *Saccharomyces cerevisiae*, researchers have overexpressed genes involved in stress responses, such as those encoding for heat shock proteins, which help stabilize proteins and protect cells under stressful conditions. Additionally, genetic modifications have been made to membrane synthesis pathways to improve the fluidity and stability of cellular membranes, allowing cells to survive in higher ethanol concentrations.

Another successful example is the engineering of *Escherichia coli* for improved tolerance to organic acids. By increasing the expression of proton pumps and buffering molecules, cells can better resist the acidity generated by high concentrations of organic acids. Other genetic engineering strategies involve modifying transport proteins to facilitate the efflux of toxic intermediates, reducing intracellular accumulation and enhancing tolerance.

8.3 ADAPTIVE LABORATORY EVOLUTION (ALE) FOR ROBUSTNESS

Adaptive Laboratory Evolution (ALE) is a powerful tool that uses the principle of natural selection to improve microbial tolerance. In ALE experiments, microorganisms are exposed to stressful conditions over multiple generations, allowing only the most tolerant cells to survive and proliferate. Over time, this process results in the accumulation of beneficial mutations that improve the organism's robustness to specific stress factors. One of the major benefits of ALE is its ability to generate novel mutations that confer tolerance, even in cases where the underlying genetic basis is unknown. For instance, ALE has been used to evolve yeast strains with high ethanol tolerance, making them more suitable for biofuel production. Similarly, ALE has produced *E. coli* strains with improved tolerance to high concentrations of succinic acid, a valuable platform chemical.

While ALE can be time-consuming, it is a highly effective strategy, especially when combined with whole-genome sequencing to identify beneficial mutations. These mutations can then be engineered into other strains or optimized further through genetic engineering, creating strains with enhanced robustness without the need for exhaustive trial and error.

8.4 SYSTEMS BIOLOGY AND OMICS FOR INSIGHT INTO STRESS TOLERANCE

8.4.1 Systems Biology

Systems biology is the computational and mathematical analysis and modeling of complex biological systems. It is a biology-based interdisciplinary field of study that focuses on complex interactions within biological systems, using a holistic approach (holism instead of the more traditional reductionism) to biological research.

Particularly from the year 2000 onwards, the concept has been used widely in biology in a variety of contexts. The Human Genome Project is an example of applied systems thinking in biology which has led to new, collaborative ways of working on problems in the biological field of genetics. One of the aims of systems biology is to model and discover emergent properties, properties of cells, tissues and organisms functioning as a system whose theoretical description is only possible using techniques of systems biology. These typically involve metabolic networks or cell signaling networks.

Systems biology can be considered from a number of different aspects. As a field of study, particularly, the study of the interactions between the components of biological systems, and how these interactions give rise to the function and behavior of that system (for example, the enzymes and metabolites in a metabolic pathway or the heart beats).

As a paradigm, systems biology is usually defined in antithesis to the so-called reductionist paradigm (biological organisation), although it is consistent with the scientific method. The distinction between the two paradigms is referred to in these quotations: "the reductionist approach has successfully identified most of the components and many of the interactions but, unfortunately, offers no convincing concepts or methods to understand how system properties emerge ... the pluralism of causes and effects in biological networks is better addressed by observing, through quantitative measures, multiple components simultaneously and by rigorous data integration with mathematical models." (Sauer *et al.*) "Systems biology ... is about putting together rather than taking apart, integration rather than reduction. It requires that we develop ways of thinking about integration that are as rigorous as our reductionist programmes, but different. ... It means changing our philosophy, in the full sense of the term." (Denis Noble)

As a series of operational protocols used for performing research, namely a cycle composed of theory, analytic or computational modelling to propose specific testable hypotheses about a biological system, experimental validation, and then using the newly acquired quantitative description of cells or cell processes to refine the computational model or theory. Since the objective is a model of the interactions in a system, the experimental techniques that most suit systems biology are those that are system-wide and attempt to be as complete as possible. Therefore, transcriptomics, metabolomics, proteomics and high-throughput techniques are used to collect quantitative data for the construction and validation of models.

As the application of dynamical systems theory to molecular biology. Indeed, the focus on the dynamics of the studied systems is the main conceptual difference between systems biology and bioinformatics. As a socioscientific phenomenon defined by the strategy of pursuing integration of complex data about the interactions in biological systems from diverse experimental sources using interdisciplinary tools and personnel.

History

Although the concept of a systems view of cellular function has been well understood since at least the 1930s, technological limitations made it difficult to make systems wide measurements. The advent of microarray technology in the 1990s opened up an entire new visa for studying cells at the systems level. In 2000, the Institute for Systems Biology was established in Seattle in an effort to lure "computational" type people who it was felt were not attracted to the academic settings of the university. The institute did not have a clear definition of what the field actually was: roughly bringing together people from diverse fields to use computers to holistically study biology in new ways. A Department of Systems Biology at Harvard Medical School was launched in 2003. In 2006 it was predicted that the buzz generated by the "very fashionable" new concept would cause all the major universities to need a systems biology department, thus that there would be careers available for graduates with a modicum of ability in computer programming and biology. In 2006 the National Science Foundation put forward a challenge to build a mathematical model of the whole cell. In 2012 the first whole-cell model of *Mycoplasma genitalium* was achieved by the Covert Laboratory at Stanford University. The whole-cell model is able to predict viability of *M. genitalium* cells in response to genetic mutations.

An earlier precursor of systems biology, as a distinct discipline, may have been by systems theorist Mihajlo Mesarovic in 1966 with an international symposium at the Case Institute of Technology in Cleveland, Ohio, titled *Systems Theory and Biology*. Mesarovic predicted that perhaps in the future there would be such a thing as "systems biology". Other early precursors that focused on the view that biology should be analyzed as a system, rather than a simple collection of parts, were Metabolic Control Analysis, developed by Henrik Kacser and Jim Burns later thoroughly revised, and Reinhart Heinrich and Tom Rapoport, and Biochemical Systems Theory developed by Michael Savageau

According to Robert Rosen in the 1960s, holistic biology had become passé by the early 20th century, as more empirical science dominated by molecular chemistry had become popular. Echoing him forty years later in 2006 Kling writes that the success of molecular biology throughout the 20th century had suppressed holistic computational methods. By 2011 the National Institutes of Health had made grant money available to support over ten systems biology centers in the United States, but by 2012 Hunter writes that systems biology still has someway to go to achieve its full potential. Nonetheless, proponents hoped that it might once prove more useful in the future.

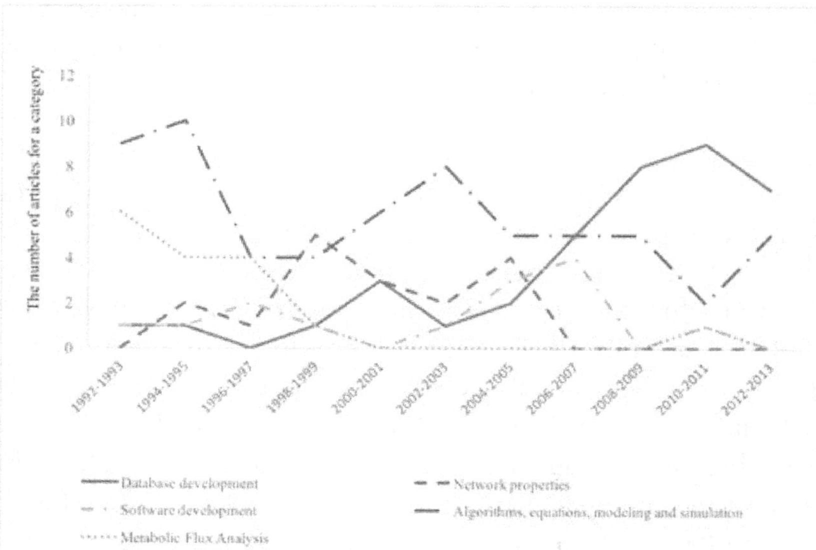

Fig 1: Shows Trends in Systems Biology Research

An important milestone in the development of systems biology has become the international project Physiome.

Associated Disciplines

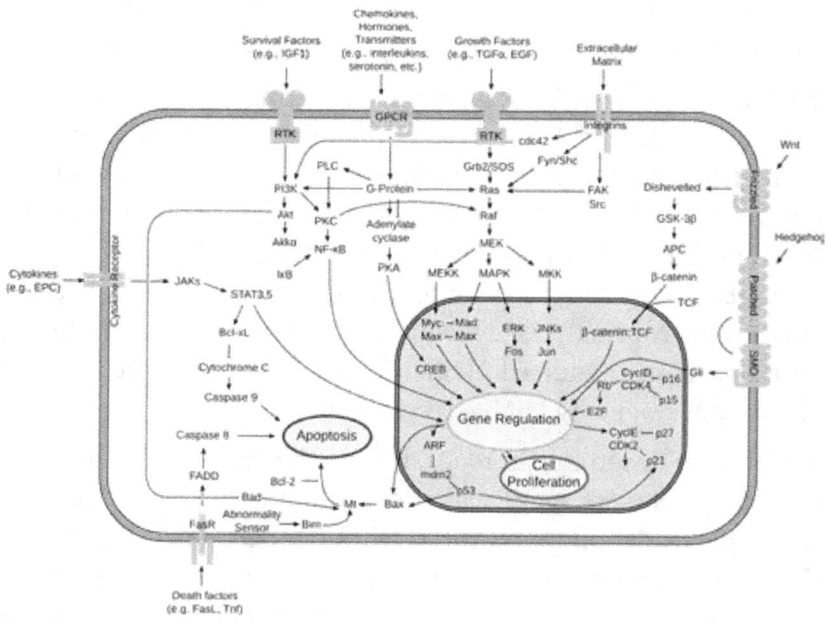

Fig 2: Overview of Signal Transduction Pathways

According to the interpretation of systems biology as using large data sets using interdisciplinary tools, a typical application is metabolomics, which is the complete set of all the metabolic products, metabolites, in the system at the organism, cell, or tissue level.

Items that may be a computer database include: phenomics, organismal variation in phenotype as it changes during its life span; genomics, organismal deoxyribonucleic acid (DNA) sequence, including intra-organismal cell specific variation. (i.e., telomere length variation); epigenomics/epigenetics, organismal and corresponding cell specific transcriptomic regulating factors not empirically coded in the genomic sequence. (i.e., DNA methylation, Histone acetylation and deacetylation, etc.); transcriptomics, organismal, tissue or whole cell gene expression measurements by DNA microarrays or serial analysis of gene expression; interferomics, organismal, tissue, or cell-level transcript correcting factors (i.e., RNA interference), proteomics, organismal, tissue, or cell level measurements of proteins and peptides via two-dimensional gel electrophoresis, mass spectrometry or multi-dimensional protein identification techniques (advanced HPLC systems coupled with mass spectrometry). Sub disciplines include phosphoproteomics, glycoproteomics and other methods to detect chemically modified proteins; glycomics, organismal, tissue, or cell-level measurements of carbohydrates; lipidomics, organismal, tissue, or cell level measurements of lipids.

The molecular interactions within the cell are also studied, this is called interactomics. A discipline in this field of study is protein–protein interactions, although interactomics includes the interactions of other molecules. Neuroelectrodynamics, where the computer's or a brain's computing function as a dynamic system is studied along with its (bio)physical mechanisms; and fluxomics, measurements of the rates of metabolic reactions in a biological system (cell, tissue, or organism).

In approaching a systems biology problem there are two main approaches. These are the top down and bottom up approach. The top down approach takes as much of the system into account as possible and relies largely on experimental results. The RNA-Seq technique is an example of an experimental top down approach. Conversely, the bottom up approach is used to create detailed models while also incorporating experimental data. An example of the bottom up approach is the use of circuit models to describe a simple gene network.

Various technologies utilized to capture dynamic changes in mRNA, proteins, and post-translational modifications. Mechanobiology, forces and physical properties at all scales, their interplay with other regulatory mechanisms; biosemiotics, analysis of the system of sign relations of an organism or other biosystems; Physiomics, a systematic study of physiome in biology.

Cancer systems biology is an example of the systems biology approach, which can be distinguished by the specific object of study (tumorigenesis and treatment of cancer). It works with the specific data (patient samples, high-throughput data with particular attention to characterizing cancer genome in patient tumour samples) and tools (immortalized cancer cell lines, mouse models of tumorigenesis, xenograft models, high-throughput sequencing methods, siRNA-based gene knocking down high-throughput screenings, computational modeling of the consequences of somatic mutations and genome instability).

The long-term objective of the systems biology of cancer is ability to better diagnose cancer, classify it and better predict the outcome of a suggested treatment, which is a basis for personalized cancer medicine and virtual cancer patient in more distant prospective. Significant efforts in computational systems biology of cancer have been made in creating realistic multi-scale *in silico* models of various tumours.

The systems biology approach often involves the development of mechanistic models, such as the reconstruction of dynamic systems from the quantitative properties of their elementary building blocks.

For instance, a cellular network can be modelled mathematically using methods coming from chemical kinetics and control theory. Due to the large number of parameters, variables and constraints in cellular networks, numerical and computational techniques are often used (e.g., flux balance analysis).

Bioinformatics and Data Analysis

Other aspects of computer science, informatics, and statistics are also used in systems biology. These include new forms of computational models, such as the use of process calculi to model biological processes (notable approaches include stochastic π-calculus, BioAmbients, Beta Binders, BioPEPA, and Brane calculus) and constraint-based modeling; integration of information from the literature, using techniques of information extraction and text mining; development of online databases and repositories for sharing data and models, approaches to database integration and software interoperability via loose coupling of software, websites and databases, or commercial suits; network-based approaches for analyzing high dimensional genomic data sets.

For example, weighted correlation network analysis is often used for identifying clusters (referred to as modules), modeling the relationship between clusters, calculating fuzzy measures of cluster (module) membership, identifying intramodular hubs, and for studying cluster preservation in other data sets; pathway-based methods for omics data analysis, e.g. approaches to identify and score pathways with differential activity of their gene, protein, or metabolite members. Much of the analysis of genomic data sets also include identifying correlations. Additionally, as much of the information comes from different fields, the development of syntactically and semantically sound ways of representing biological models is needed.

Creating Biological Models

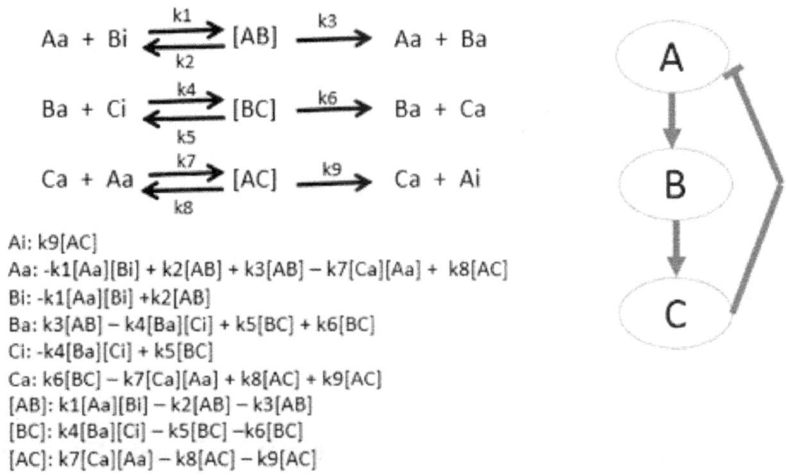

Fig 3: A Simple Three Protein Negative Feedback Loop Modeled with Mass Action Kinetic Differential Equations

Researchers begin by choosing a biological pathway and diagramming all of the protein, gene, and/or metabolic pathways. After determining all of the interactions, mass action kinetics or enzyme kinetic rate laws are used to describe the speed of the reactions in the system. Using mass-conservation, the differential equations for the biological system can be constructed. Experiments or parameter fitting can be done to determine the parameter values to use in the differential equations. These parameter values will be the various kinetic constants required to fully describe the model. This model determines the behavior of species in biological systems and bring new insight to the specific activities of system. Sometimes it is not possible to gather all reaction rates of a system. Unknown reaction rates are determined by simulating the model of known parameters and target behavior which provides possible parameter values.

The use of constraint-based reconstruction and analysis (COBRA) methods has become popular among systems biologists to simulate and predict the metabolic phenotypes, using genome-scale models. One of the methods is the flux balance analysis (FBA) approach, by which one can study the biochemical networks and analyze the flow of metabolites through a particular metabolic network, by optimizing the objective function of interest (e.g. maximizing biomass production to predict growth).

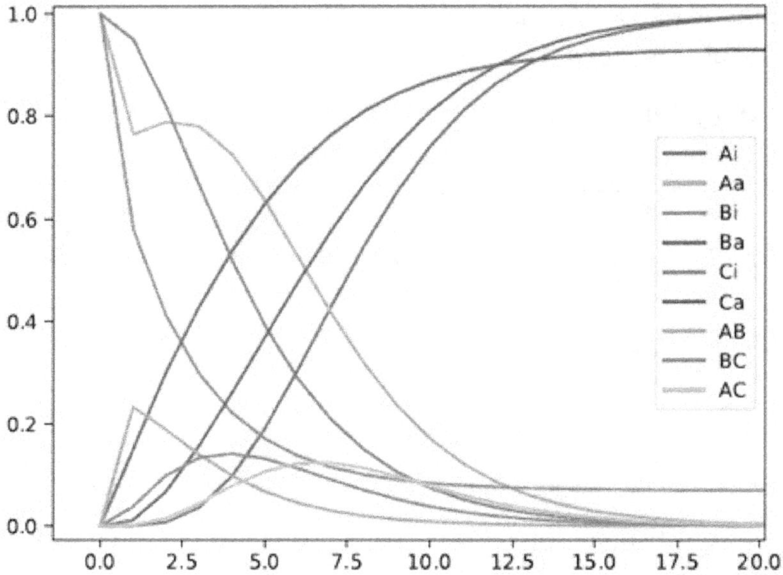

Fig 4: Plot of Concentrations vs. Time for the Simple Three Protein Negative Feedback Loop

All parameters are set to either 0 or 1 for initial conditions. The reaction is allowed to proceed until it hits equilibrium. This plot is of the change in each protein over time.

8.4.2 Omics

The branches of science known informally as omics are various disciplines in biology whose names end in the suffix *-omics*, such as genomics, proteomics, metabolomics, metagenomics, phenomics and transcriptomics. Omics aims at the collective characterization and quantification of pools of biological molecules that translate into the structure, function, and dynamics of an organism or organisms.

The related suffix -ome is used to address the objects of study of such fields, such as the genome, proteome or metabolome respectively.

Functional genomics aims at identifying the functions of as many genes as possible of a given organism. It combines different -omics techniques such as transcriptomics and proteomics with saturated mutant collections.

Origin

The *Oxford English Dictionary* (*OED*) distinguishes three different fields of application for the *-ome* suffix:

- In medicine, forming nouns with the sense "swelling, tumour",
- In botany or zoology, forming nouns in the sense "a part of an animal or plant with a specified structure", and

- In cellular and molecular biology, forming nouns with the sense "all constituents considered collectively".

Bioinformaticians and molecular biologists figured amongst the first scientists to apply the "-ome" suffix widely. Early advocates included bioinformaticians in Cambridge, UK, where there were many early bioinformatics labs such as the MRC centre, Sanger centre, and EBI (European Bioinformatics Institute); for example, the MRC centre carried out the first genome and proteome projects.

Current Usage

Many "omes" beyond the original "genome" have become useful and have been widely adopted by research scientists. "Proteomics" has become well-established as a term for studying proteins at a large scale. "Omes" can provide an easy shorthand to encapsulate a field; for example, an interactomics study is clearly recognisable as relating to large-scale analyses of gene-gene, protein-protein, or protein-ligand interactions. Researchers are rapidly taking up omes and omics, as shown by the explosion of the use of these terms in PubMed since the mid-1990s.

8.4.3 Exploring Stress Tolerance Through Systems Biology and Omics

Systems biology, an approach that studies the interactions within biological systems, has become invaluable for understanding the complex responses of microorganisms to stress. Omics technologies, including genomics, transcriptomics, proteomics, and metabolomics, allow scientists to analyze the effects of stress on various cellular pathways and identify key regulatory networks involved in stress tolerance. For instance, transcriptomic studies can reveal which genes are upregulated or downregulated in response to specific stressors, such as high ethanol concentrations. Proteomic analysis provides insight into changes in protein abundance and modifications, helping to identify stress-response proteins. By integrating data from multiple omics approaches, researchers can construct models of cellular stress responses, which can then guide genetic engineering strategies to enhance tolerance.

One example of a systems biology approach is the use of flux balance analysis (FBA), which models metabolic fluxes to predict how cells respond to stress. By identifying bottlenecks in metabolic pathways under stress conditions, scientists can target specific enzymes or pathways for genetic modification. This systems-level understanding enables more precise engineering, minimizing unintended consequences and improving overall strain performance.

8.5 METABOLIC ENGINEERING AND PATHWAY OPTIMIZATION

8.5.1 Metabolic Engineering

Metabolic engineering is the practice of optimizing genetic and regulatory processes within cells to increase the cell's production of a certain substance. These processes are chemical networks that use a series of biochemical reactions and enzymes that allow cells to convert raw materials into molecules necessary for the cell's survival. Metabolic engineering specifically seeks to mathematically model these networks, calculate a yield of useful products, and pin point parts of the network that constrain the production of these products. Genetic engineering techniques can then be used to modify the network in order to relieve these constraints. Once again this modified network can be modeled to calculate the new product yield.

The ultimate goal of metabolic engineering is to be able to use these organisms to produce valuable substances on an industrial scale in a cost-effective manner. Current examples include producing beer, wine, cheese, pharmaceuticals, and other biotechnology products. Another possible area of use is the development of oil crops whose composition has been modified to improve their nutritional value. Some of the common strategies used for metabolic engineering are:

- Overexpressing the gene encoding the rate-limiting enzyme of the biosynthetic pathway.
- Blocking the competing metabolic pathways.
- Heterologous gene expression.
- Enzyme engineering.

Since cells use these metabolic networks for their survival, changes can have drastic effects on the cells' viability. Therefore, trade-offs in metabolic engineering arise between the cells ability to produce the desired substance and its natural survival needs. Therefore, instead of directly deleting and/or overexpressing the genes that encode for metabolic enzymes, the current focus is to target the regulatory networks in a cell to efficiently engineer the metabolism.

History and Applications

In the past, to increase the productivity of a desired metabolite, a microorganism was genetically modified by chemically induced mutation, and the mutant strain that overexpressed the desired metabolite was then chosen. However, one of the main problems with this technique was that the metabolic pathway for the production of that metabolite was not analyzed, and as a result, the constraints to production and relevant pathway enzymes to be modified were unknown.

In 1990s, a new technique called metabolic engineering emerged. This technique analyzes the metabolic pathway of a microorganism, and determines the constraints and their effects on the production of desired compounds. It then uses genetic engineering to relieve these constraints. Some examples of successful metabolic engineering are the following: (i) Identification of constraints to lysine production in *Corynebacterium glutamicum* and insertion of new genes to relieve these constraints to improve production (ii) Engineering of a new fatty acid biosynthesis pathway, called reversed beta oxidation pathway, that is more efficient than the native pathway in producing fatty acids and alcohols which can potentially be catalytically converted to chemicals and fuels (iii) Improved production of DAHP an aromatic metabolite produced by *E. coli* that is an intermediate in the production of aromatic amino acids. It was determined through metabolic flux analysis that the theoretical maximal yield of DAHP per glucose molecule utilized, was 3/7. This is because some of the carbon from glucose is lost as carbon dioxide, instead of being utilized to produce DAHP. Also, one of the metabolites (PEP, or phosphoenolpyruvate) that are used to produce DAHP, was being converted to pyruvate (PYR) to transport glucose into the cell, and therefore, was no longer available to produce DAHP. In order to relieve the shortage of PEP and increase yield, Patnaik et al. used genetic engineering on *E. coli* to introduce a reaction that converts PYR back to PEP. Thus, the PEP used to transport glucose into the cell is regenerated, and can be used to make DAHP. This resulted in a new theoretical maximal yield of 6/7 – double that of the native *E. coli* system.

At the industrial scale, metabolic engineering is becoming more convenient and cost-effective. According to the Biotechnology Industry Organization, "more than 50 biorefinery facilities are being built across North America to apply metabolic engineering to produce biofuels and chemicals from renewable biomass which can help reduce greenhouse gas emissions". Potential biofuels include short-chain alcohols and alkanes (to replace gasoline), fatty acid methyl esters and fatty alcohols (to replace diesel), and fatty acid-and isoprenoid-based biofuels (to replace diesel).

Metabolic engineering continues to evolve in efficiency and processes aided by breakthroughs in the field of synthetic biology and progress in understanding metabolite damage and its repair or preemption. Early metabolic engineering experiments showed that accumulation of reactive intermediates can limit flux in engineered pathways and be deleterious to host cells if matching damage control systems are missing or inadequate. Researchers in synthetic biology optimize genetic pathways, which in turn influence cellular metabolic outputs. Recent decreases in cost of synthesized DNA and developments in genetic circuits help to influence the ability of metabolic engineering to produce desired outputs.

Metabolic Flux Analysis

Setting up a Metabolic Pathway for Analysis

The first step in the process is to identify a desired goal to achieve through the improvement or modification of an organism's metabolism. Reference books and online databases are used to research reactions and metabolic pathways that are able to produce this product or result. These databases contain copious genomic and chemical information including pathways for metabolism and other cellular processes. Using this research, an organism is chosen that will be used to create the desired product or result. Considerations that are taken into account when making this decision are how close the organism's metabolic pathway is to the desired pathway, the maintenance costs associated with the organism, and how easy it is to modify the pathway of the organism. *Escherichia coli* (*E. coli*) is widely used in metabolic engineering to synthesize a wide variety of products such as amino acids because it is relatively easy to maintain and modify. If the organism does not contain the complete pathway for the desired product or result, then genes that produce the missing enzymes must be incorporated into the organism.

Analyzing a Metabolic Pathway

The completed metabolic pathway is modeled mathematically to find the theoretical yield of the product or the reaction fluxes in the cell. A flux is the rate at which a given reaction in the network occurs.

Simple metabolic pathway analysis can be done by hand, but most require the use of software to perform the computations. These programs use complex linear algebra algorithms to solve these models. To solve a network using the equation for determined systems shown below, one must input the necessary information about the relevant reactions and their fluxes. Information about the reaction (such as the reactants and stoichiometry) are contained in the matrices G_x and G_m. Matrices V_m and V_x contain the fluxes of the relevant reactions. When solved, the equation yields the values of all the unknown fluxes (contained in V_x).

$$V_x = -(G_x)^{-1} * (G_m * V_m)$$

Determining the Optimal Genetic Manipulations

After solving for the fluxes of reactions in the network, it is necessary to determine which reactions may be altered in order to maximize the yield of the desired product.

To determine what specific genetic manipulations to perform, it is necessary to use computational algorithms, such as OptGene or OptFlux. They provide recommendations for which genes should be overexpressed, knocked out, or introduced in a cell to allow increased production of the desired product. For example, if a given reaction has particularly low flux and is limiting the amount of product, the software may recommend that the enzyme catalyzing this reaction should be overexpressed in the cell to increase the reaction flux. The necessary genetic manipulations can be performed using standard molecular biology techniques. Genes may be overexpressed or knocked out from an organism, depending on their effect on the pathway and the ultimate goal.

Experimental Measurements

In order to create a solvable model, it is often necessary to have certain fluxes already known or experimentally measured. In addition, in order to verify the effect of genetic manipulations on the metabolic network (to ensure they align with the model), it is necessary to experimentally measure the fluxes in the network. To measure reaction fluxes, carbon flux measurements are made using carbon-13 isotopic labeling. The organism is fed a mixture that contains molecules where specific carbons are engineered to be carbon-13 atoms, instead of carbon-12. After these molecules are used in the network, downstream metabolites also become labeled with carbon-13, as they incorporate those atoms in their structures. The specific labeling pattern of the various metabolites is determined by the reaction fluxes in the network. Labeling patterns may be measured using techniques such as gas chromatography-mass spectrometry (GC-MS) along with computational algorithms to determine reaction fluxes.

8.5.2 Optimizing Pathways Through Metabolic Engineering

Metabolic engineering involves the optimization of metabolic pathways to improve the efficiency of microbial production processes.

This approach can also enhance stress tolerance by redirecting metabolic fluxes away from pathways that produce toxic intermediates or by strengthening pathways involved in stress response.

For example, in the production of biofuels, researchers have rerouted metabolic pathways in *E. coli* to limit the production of reactive oxygen species (ROS), which are toxic byproducts of cellular metabolism. By introducing pathways that neutralize ROS, engineered strains exhibit higher tolerance to high-product conditions, such as those seen in biofuel fermentation.

Additionally, metabolic engineering can optimize energy production and resource allocation within cells, ensuring that adequate energy is available to activate stress response pathways.

This approach has been applied to increase temperature tolerance in thermophilic bacteria, which are used in high-temperature industrial processes. By optimizing energy production pathways, cells can better withstand the high-energy demands of heat shock responses, resulting in strains that are more resilient under extreme conditions.

8.6 FUTURE PROSPECTS: AI AND MACHINE LEARNING IN STRAIN ENGINEERING

Looking forward, artificial intelligence (AI) and machine learning (ML) are expected to play a major role in enhancing the tolerance and robustness of industrial microorganisms. By analyzing large omics datasets, AI algorithms can identify patterns and correlations between genetic factors and stress tolerance, enabling researchers to make informed decisions about which genes or pathways to target.

Machine learning models can predict the effects of genetic modifications on cell behavior, allowing for rapid strain design and testing. This capability is particularly useful for identifying tolerance mechanisms that are difficult to study through conventional methods, such as the interaction between multiple genes or the effects of complex metabolic pathways.

In addition, AI-driven approaches could facilitate the development of adaptive metabolic control systems.

These systems would use biosensors to detect stress levels in real-time and adjust gene expression accordingly, allowing cells to dynamically respond to changing environmental conditions. Such "smart" strains would automatically activate stress response pathways when needed, conserving energy and improving overall robustness.

8.7 EMERGING TECHNOLOGIES: CRISPR AND SYNTHETIC BIOLOGY

CRISPR technology has revolutionized the field of genetic engineering, enabling precise and efficient modifications to microbial genomes. This precision is invaluable for enhancing tolerance, as researchers can edit specific genes or regulatory elements involved in stress response with high accuracy. By using CRISPR, scientists have developed yeast strains with enhanced resistance to organic acids and bacteria with improved solvent tolerance.

In addition, synthetic biology offers new approaches for constructing synthetic pathways and regulatory networks tailored to specific industrial needs. Synthetic pathways designed for tolerance can be added to microbial genomes, allowing cells to handle high concentrations of toxic products or intermediates.

For instance, researchers have engineered synthetic pathways to convert toxic intermediates into less harmful compounds, improving the viability of cells under high-product conditions.

8.8 CONCLUSION

Increasing the tolerance and robustness of industrial microorganisms is essential for maximizing productivity and sustainability in bioprocesses. Current strategies, such as genetic engineering, ALE, and systems biology, have greatly expanded the capabilities of microbial cell factories, enabling them to withstand harsh industrial environments. Advances in metabolic engineering, pathway optimization, and emerging technologies such as CRISPR and AI are driving further improvements, allowing researchers to design highly tolerant and robust strains.

In the future, the integration of AI, synthetic biology, and precision engineering tools will enable the creation of microbial strains that are both highly productive and resilient. As tolerance and robustness improvements continue to evolve, industrial biotechnology will become more efficient and versatile, with applications extending to a broader range of products and sustainable processes. These advances hold significant promise for meeting the demands of modern industry while minimizing the environmental footprint of microbial production systems, ultimately supporting a more sustainable and bio-based economy.

CHAPTER 9

ENGINEERING MICROORGANISMS FOR THE PRODUCTION OF PHARMACEUTICAL BIOMOLECULES: CURRENT TRENDS AND FUTURE PROSPECTS

Microorganisms have long been valuable tools in biotechnology, especially for the production of pharmaceutical biomolecules, including antibiotics, enzymes, vaccines, and therapeutic proteins.

Advances in genetic engineering, synthetic biology, and metabolic engineering have expanded the potential of microbes, such as bacteria, yeast, and fungi, to produce complex, high-value pharmaceutical products. By optimizing microbial strains, researchers can improve yields, lower costs, and reduce dependency on more resource-intensive production methods, such as those involving mammalian cells.

9.1 THE ROLE OF MICROORGANISMS IN PHARMACEUTICAL PRODUCTION

Microorganisms are naturally equipped to produce a wide range of biomolecules that have therapeutic applications. For instance, bacteria like *Escherichia coli* and fungi like *Aspergillus niger* have been used to produce antibiotics and enzymes, while yeast such as *Saccharomyces cerevisiae* has been a workhorse for recombinant protein production. Microorganisms are advantageous in pharmaceutical production due to their rapid growth rates, ease of genetic manipulation, and ability to thrive in large-scale fermentation systems.

Over recent decades, microbial production has been applied to produce not only small molecules, such as antibiotics, but also complex molecules like insulin and monoclonal antibodies.

As the demand for biologics increases, microbial production platforms offer an attractive alternative to traditional methods, enabling the efficient synthesis of pharmaceutical compounds that would be difficult or expensive to produce through chemical synthesis or animal cell culture systems.

9.2 GENETIC ENGINEERING FOR ENHANCED PHARMACEUTICAL PRODUCTION

Genetic engineering is at the heart of microbial strain development for pharmaceutical production. Through precise modifications to the microbial genome, scientists can optimize pathways involved in the biosynthesis of specific biomolecules. Techniques such as CRISPR-Cas9 and other gene-editing tools have made it possible to engineer microorganisms more efficiently, allowing for targeted modifications with minimal off-target effects.

One prominent example is the production of insulin in *E. coli*. Initially, *E. coli* was genetically modified to produce human insulin, revolutionizing the treatment of diabetes by providing a more sustainable and accessible supply of the hormone. Similarly, yeasts are often engineered to produce therapeutic proteins, including hepatitis B vaccines and human growth hormone, due to their capacity to perform post-translational modifications, which are crucial for the functionality of many proteins.

In addition to pathway optimization, genetic engineering can enhance microbial tolerance to the product being synthesized, thereby increasing production efficiency. By engineering transporter proteins and stress-response genes, researchers have improved the yield of antibiotics and other biomolecules in microbial systems. For instance, tolerance to high concentrations of penicillin in *Penicillium chrysogenum* has been enhanced through genetic modifications, allowing for greater antibiotic production.

9.3 SYNTHETIC BIOLOGY AND METABOLIC ENGINEERING

Synthetic biology has transformed the way pharmaceutical compounds are produced by enabling the design of synthetic pathways in microorganisms. In synthetic biology, modular gene circuits are introduced into microbes to control the synthesis of pharmaceutical compounds with high precision. For example, engineered bacteria have been used to produce complex molecules like artemisinin, an antimalarial compound that was previously extracted from plants at great expense and limited availability. By introducing a synthetic pathway into *Saccharomyces cerevisiae*, scientists were able to produce artemisinin in yeast, ensuring a more reliable supply.

Metabolic engineering is closely linked to synthetic biology and involves optimizing entire metabolic pathways to increase the flow of precursors and intermediates needed for pharmaceutical production. Through metabolic engineering, researchers can improve the flux toward desired compounds and reduce the formation of unwanted byproducts. For instance, by modifying pathways related to amino acid biosynthesis, scientists have engineered *E. coli* to produce a variety of amino acid-based drugs, such as D-alanine, which is used in the synthesis of antibiotics and immunosuppressants.

9.4 MICROBIAL CELL FACTORIES FOR THE PRODUCTION OF BIOLOGICS

Microbial cell factories have become essential for the production of biologics, a class of drugs that includes therapeutic proteins, enzymes, and monoclonal antibodies. These complex molecules often require post-translational modifications, such as glycosylation, which bacteria do not naturally perform. However, advances in protein engineering have enabled researchers to engineer microbial hosts capable of such modifications, allowing bacteria and yeast to produce more human-like proteins.

Prior to MCFs, scientists employed traditional engineering techniques to produce various commodities. These methodologies include modifying metabolic pathways, eliminating enzymes, or the balancing of ATP to drive metabolic flux. However, when these approaches were applied for industrial productions, they could not withstand the industrial environments that consisted of toxins and fluctuating temperatures. Ultimately, the techniques were never able to scale up and output bio-products that were obtained in the laboratory.

Thus, MCFs were developed by using a heterogenous biosynthesis pathway in a microbial host. As a host, MCFs take in various substrates and convert them into valuable compounds. These products can range from fuels, chemical, food ingredients, to pharmaceuticals.

In microbial cells, the cell walls are either Gram-positive or Gram-negative. These outcomes are based on the Gram Stain test. Gram-positive cell walls have thick peptidoglycan layer and no outer lipid membrane while Gram-negative bacteria have a thin peptidoglycan layer and an outer lipid membrane. Although a thick Gram-positive cell wall is advantageous, it is easier to attack as the peptidoglycan layer absorbs antibiotics and cleaning products. A Gram-negative cell wall is more resistant to such attacks and more difficult to destroy.

The membrane of microbial cells are bilayers, composed of phospholipids. The phospholipids may range in chain length to branching. Ultimately, the phospholipid will determine the membrane properties, such as fluidity and charge, that will regulate the interactions with nearby proteins. In addition, the membrane oversees the development of the cell's morphology and cell sizes. *Escherichia coli* is often utilized a base line to differentiate and define the membrane of MCFs.

The nucleoid forms an irregular shaped region within a prokaryote cell, containing all or majority of the genetic material to reproduce. The nucleoid controls the activity of the MCF and reproduction of itself and products.

Current methods of programming MCFs utilize strain engineering, which rely on random mutagenesis. In addition, the conventional techniques are labor-intensive, timely, and difficult to analyze. This has led many scientific trials to utilize genomic editing tools to improve MCFs, such as ZFNs, TALENs, and CRISPR. These approaches allow genetic manipulation and analysis, specifically creating double stranded breaks within a genome sequence.

Zinc-finger nucleases (ZFNs) were the first genomic editing tool to be able to target any genomic site. By inducing a double-stranded break, ZFNs can facilitate targeted editing. However, when employed to reinforce MCFs, ZFNs have an unusual low success rate. In various trials, the ZFNs were unable to obtain a three-finger array or the triplet was unable to be assembled into a new sequence. Thus, incorporation of ZFNs into MCFs has remained strenuous and costly.

Transcription activator-like effector nucleases (TALENs) work in a similar manner to ZFNs, but TALENs are based on fusion proteins. TALENs have been applied to numerous MCFs, such as yeast and zebrafish. Many developments has explored fairyTALE, a liquid phase synthesis TALEN platform, to create nucleases, activators, and repressors for MCFs. Although TALENs have fewer obstacles than ZFNs, they are still troublesome as assembling large quantities of repeats into an array remains a significant problem.

Clustered regularly interspaced palindromic repeats (CRISPR) and its associated proteins (Cas) has become one of the most popular genome editing tools due to its efficiency and low cost. The CRISPR/CAS9 has been utilized to enhance MCFs to produce yeast, bacteria, and E.coli. When optimizing yeast, CRISPR/CAS9 promoting S.pyogenes has been found to be the most influential strategy. For E.coli, studies have determined a strategy preventing genome instability to be the most robust metabolic engineering approach regardless of the specific methodology.

The most significant advantage of MCFs is the ability to be utilized in industrial environments with minimal limitations. Through metabolic engineering, MCFs rely on innovative strategic tools for the development and optimization of metabolic and gene regulatory networks for efficient production. Going from lab to large scale development involves consideration of three factors: product yield, productivity, and the product titre. A common dilemma however is the trade-off between product yield and productivity. If a company maximizes productivity, they will ultimately lower their product yield and vice versa.

To combat this issue, strategies have been developed to maximize all three factors. One of the most common techniques is utilizing fed-batch culture. Fed-batch culture is, in the broadest sense, defined as an operational technique in biotechnological processes where one or more nutrients (substrates) are fed (supplied) to the bioreactor during cultivation and in which the product(s) remain in the bioreactor until the end of the run. Another method is utilizing continuous cultivation strategy. The premise behind continuous cultivation is to maintain a steady-state cell metabolism over long periods of times. By having multiple approaches for MCF, companies may customize each process to their specific product(s).

Yeast, particularly *Pichia pastoris* and *Saccharomyces cerevisiae*, has been a popular choice for producing therapeutic proteins because it combines the growth advantages of microorganisms with the ability to perform some post-translational modifications.

Researchers have engineered yeast strains to mimic human glycosylation patterns, making it possible to produce glycosylated proteins, such as antibodies and cytokines, for pharmaceutical applications.

Bacteria, too, are being engineered to produce biologics that were once thought to be possible only in mammalian cells. Using glycoengineering, researchers have introduced genes responsible for human-like glycosylation into *E. coli*, creating bacterial strains that can produce functional antibodies. Such innovations hold promise for reducing the cost and complexity of producing biologics, making these important drugs more accessible to a wider patient population.

9.5 THE ROLE OF HIGH-THROUGHPUT SCREENING AND DIRECTED EVOLUTION

High-throughput screening and directed evolution are critical techniques used to optimize microbial strains for pharmaceutical production. In directed evolution, microbial populations are subjected to multiple rounds of mutation and selection under conditions that favor enhanced production of the desired compound.

High-throughput screening, on the other hand, enables the rapid identification of strains with desirable traits by testing thousands of microbial variants in parallel.

These techniques have been especially useful in optimizing enzyme function for pharmaceutical synthesis. Enzymes are frequently used as biocatalysts to produce chiral drugs, which are more effective in their active form.

By applying directed evolution and high-throughput screening, researchers have developed microbial strains that produce highly selective enzymes, allowing for the synthesis of chiral drugs with greater efficiency and fewer side products.

For example, directed evolution has been used to improve the activity of cytochrome P450 enzymes in *E. coli*, enabling the microbial production of steroids, which are used in a variety of pharmaceutical applications.

By evolving enzymes to achieve higher specificity and efficiency, the production of these complex biomolecules becomes feasible on an industrial scale.

9.6 FUTURE PROSPECTS: AI AND MACHINE LEARNING IN STRAIN ENGINEERING

9.6.1 Understanding the Strain Engineering

Strain engineering refers to a general strategy employed in semiconductor manufacturing to enhance device performance. Performance benefits are achieved by modulating strain, as one example, in the transistor channel, which enhances electron mobility (or hole mobility) and thereby conductivity through the channel. Another example are semiconductor photocatalysts strain-engineered for more effective use of sunlight.

In CMOS Manufacturing

The use of various strain engineering techniques has been reported by many prominent microprocessor manufacturers, including AMD, IBM, and Intel, primarily with regards to sub-130 nm technologies. One key consideration in using strain engineering in CMOS technologies is that PMOS and NMOS respond differently to different types of strain. Specifically, PMOS performance is best served by applying compressive strain to the channel, whereas NMOS receives benefit from tensile strain. Many approaches to strain engineering induce strain locally, allowing both n-channel and p-channel strain to be modulated independently.

One prominent approach involves the use of a strain-inducing capping layer. CVD silicon nitride is a common choice for a strained capping layer, in that the magnitude and type of strain (e.g. tensile vs compressive) may be adjusted by modulating the deposition conditions, especially temperature. Standard lithography patterning techniques can be used to selectively deposit strain-inducing capping layers, to deposit a compressive film over only the PMOS, for example.

Capping layers are key to the *Dual Stress Liner* (DSL) approach reported by IBM-AMD. In the DSL process, standard patterning and lithography techniques are used to selectively deposit a tensile silicon nitride film over the NMOS and a compressive silicon nitride film over the PMOS.

A second prominent approach involves the use of a silicon-rich solid solution, especially silicon-germanium, to modulate channel strain. One manufacturing method involves epitaxial growth of silicon on top of a relaxed silicon-germanium underlayer. Tensile strain is induced in the silicon as the lattice of the silicon layer is stretched to mimic the larger lattice constant of the underlying silicon-germanium. Conversely, compressive strain could be induced by using a solid solution with a smaller lattice constant, such as silicon-carbon.

In Thin Films

Strain can be induced in thin films with either epitaxial growth, or more recently, topological growth. Epitaxial strain in thin films generally arises due to lattice mismatch between the film and its substrate and triple junction restructuring at the surface triple junction, which arises either during film growth or due to thermal expansion mismatch. Tuning this epitaxial strain can be used to moderate the properties of thin films and induce phase transitions. The misfit parameter (*f*) is given by the equation below:

$$f = (a_s - a_e)/a_e$$

After some critical film thickness, it becomes energetically favorable to relieve some mismatch strain through the formation of misfit dislocations or microtwins. Misfit dislocations can be interpreted as a dangling bond at an interface between layers with different lattice constants. This critical thickness (h_c) was computed by Mathews and Blakeslee to be:

$$h_c = \frac{b(2 - v cos^2\alpha)[ln(h_c / b) + 1]}{8\pi |f| (1 + v) cos\lambda}$$

The equilibrium in-plane strain for a thin film with a thickness (h) that exceeds h_c is then given by the expression:

$$\epsilon_{\|} = \frac{f}{|f|} \frac{b(1 - v cos^2(\alpha)[ln(h / b) + 1]}{8\pi |f| (1 + v) cos\lambda}$$

Strain relaxation at thin film interfaces via misfit dislocation nucleation and multiplication occurs in three stages which are distinguishable based on the relaxation rate. The first stage is dominated by glide of pre-existing dislocations and is characterized by a slow relaxation rate. The second stage has a faster relaxation rate, which depends on the mechanisms for dislocation nucleation in the material. Finally, the last stage represents a saturation in strain relaxation due to strain hardening.

Strain engineering has been well-studied in complex oxide systems, in which epitaxial strain can strongly influence the coupling between the spin, charge, and orbital degrees of freedom, and thereby impact the electrical and magnetic properties. Epitaxial strain has been shown to induce metal-insulator transitions and shift the Curie temperature for the antiferromagnetic-to-ferromagnetic transition in $La_{1-x}Sr_xMnO_3$. In alloy thin films, epitaxial strain has been observed to impact the spinodal instability, and therefore impact the driving force for phase separation. This is explained as a coupling between the imposed epitaxial strain and the system's composition-dependent elastic properties.

Researchers more recently have achieved strain in thick oxide films larger than that achieved in epitaxial growth by incorporating nano-structured topologies (Guerra and Vezenov, 2002) and nanorods/nanopillars within an oxide film matrix. Following this work, researchers world-wide have created such self-organized, phase-separated, nanorod/nanopillar structures in numerous oxide films as reviewed here. In 2008, Thulin and Guerra published calculations of strain-modified anatase titania band structures, which included an indicated higher hole mobility with increasing strain. Additionally, in two dimensional materials such as WSe_2 strain has been shown to induce conversion from an indirect semiconductor to a direct semiconductor allowing a hundred-fold increase in the light emission rate.

In III-N LEDs

Strain engineering plays a major role in III-N LEDs, one of the most ubiquitous and efficient LED varieties that has only gained popularity after the 2014 Nobel Prize in Physics. Most III-N LEDs utilize a combination of GaN and InGaN, the latter being used as the quantum well region. The composition of In within the InGaN layer can be tuned to change the color of the light emitted from these LEDs. However, the epilayers of the LED quantum well have inherently mismatched lattice constants, creating strain between the layers. Due to the quantum confined Stark effect (QCSE), the electron and hole wave functions are misaligned within the quantum well, resulting in a reduced overlap integral, decreased recombination probability, and increased carrier lifetime.

As such, applying an external strain can negate the internal quantum well strain, reducing the carrier lifetime and making the LEDs a more attractive light source for communications and other applications requiring fast modulation speeds.

With appropriate strain engineering, it is possible to grow III-N LEDs on Si substrates. This can be accomplished via strain relaxed templates, superlattices, and pseudo-substrates. Furthermore, electro-plated metal substrates have also shown promise in applying an external counterbalancing strain to increase the overall LED efficiency.

In DUV LEDs

In addition to traditional strain engineering that takes place with III-N LEDs, Deep Ultraviolet (DUV) LEDs, which use AlN, AlGaN, and GaN, undergo a polarity switch from TE to TM at a critical Al composition within the active region. The polarity switch arises from the negative value of AlN's crystal field splitting, which results in its valence bands switching character at this critical Al composition. Studies have established a linear relationship between this critical composition within the active layer and the Al composition used in the substrate templating region, underscoring the importance of strain engineering in the character of light emitted from DUV LEDs. Furthermore, any existing lattice mismatch causes phase separation and surface roughness, in addition to creating dislocations and point defects. The former results in local current leakage while the latter enhances the nonradiative recombination process, both reducing the device's internal quantum efficiency (IQE). Active layer thickness can trigger the bending and annihilation of threading dislocations, surface roughening, phase separation, misfit dislocation formation, and point defects. All of these mechanisms compete across different thicknesses. By delaying strain accumulation to grow at a thicker epilayer before reaching the target relaxation degree, certain adverse effects can be reduced.

In Nano-Scale Materials

Typically, the maximum elastic strain achievable in normal bulk materials ranges from 0.1% to 1%. This limits our ability to effectively modify material properties in a reversible and quantitative manner using strain. However, recent research on nanoscale materials has shown that the elastic strain range is much broader. Even the hardest material in nature, diamond, exhibits up to 9.0% uniform elastic strain at the nanoscale. Keeping in line with Moore's law, semiconductor devices are continuously shrinking in size to the nanoscale. With the concept of "smaller is stronger", elastic strain engineering can be fully exploited at the nanoscale.

In nanoscale elastic strain engineering, the crystallographic direction plays a crucial role. Most materials are anisotropic, meaning their properties vary with direction. This is particularly true in elastic strain engineering, as applying strain in different crystallographic directions can have a significant impact on the material's properties.

Taking diamond as an example, Density Functional Theory (DFT) simulations demonstrate distinct behaviors in the bandgap decreasing rates when strained along different directions. Straining along the <110> direction results in a higher bandgap decreasing rate, while straining along the <111> direction leads to a lower bandgap decreasing rate but a transition from an indirect to a direct bandgap. A similar indirect-direct bandgap transition can be observed in strained silicon. Theoretically, achieving this indirect-direct bandgap transition in silicon requires a strain of more than 14% uniaxial strain.

In 2D Materials

In the case of elastic strain, when the limit is exceeded, plastic deformation occurs due to slip and dislocation movement in the microstructure of the material. Plastic deformation is not commonly utilized in strain engineering due to the difficulty in controlling its uniform outcome. Plastic deformation is more influenced by local distortion rather than the global stress field observed in elastic strain. However, 2D materials have a greater range of elastic strain compared to bulk materials because they lack typical plastic deformation mechanisms like slip and dislocation. Additionally, it is easier to apply strain along a specific crystallographic direction in 2D materials compared to bulk materials.

Recent research has shown significant progress in strain engineering in 2D materials through techniques such as deforming the substrate, inducing material rippling, and creating lattice asymmetry. These methods of applying strain effectively enhance the electric, magnetic, thermal, and optical properties of the material. For example, in the reference provided, the optical gap of monolayer and bilayer MoS2 decreases at rates of approximately 45 and 120 meV/%, respectively, under 0-2.2% uniaxial strain. Additionally, the photoluminescence intensity of monolayer MoS2 decreases at 1% strain, indicating an indirect-to-direct bandgap transition. The reference also demonstrates that strain-engineered rippling in black phosphorus leads to bandgap variations between +10% and -30%. In the case of ReSe2, the literature shows the formation of local wrinkle structures when the substrate is relaxed after stretching. This folding process results in a redshift in the absorption spectrum peak, leading to increased light absorption and changes in magnetic properties and bandgap. The research team also conducted I-V curve tests on the stretched samples and found that a 30% stretching resulted in lower resistance compared to the unstretched samples. However, a 50% stretching showed the opposite effect, with higher resistance compared to the unstretched samples. This behavior can be attributed to the folding of ReSe2, with the folded regions being particularly weak.

9.6.2 AI and Machine Learning in Strain Engineering

Artificial intelligence (AI) and machine learning (ML) are emerging as powerful tools in microbial strain engineering for pharmaceutical production. By analyzing large omics datasets, including genomics, transcriptomics, and metabolomics data, AI algorithms can identify patterns that help predict the effects of genetic modifications on microbial production capacity. This capability significantly accelerates the design-build-test cycle in strain development.

ML models can also optimize metabolic pathways by predicting the most effective gene modifications to increase product yields. For example, AI-driven approaches can help identify regulatory elements that control pathway fluxes, enabling researchers to fine-tune production pathways with unprecedented precision. With the integration of AI, researchers can design microbial strains that are more robust and productive, reducing the time and resources needed for optimization.

9.7 ENVIRONMENTAL AND SUSTAINABILITY BENEFITS OF MICROBIAL PHARMACEUTICAL PRODUCTION

Microbial production of pharmaceutical biomolecules offers significant environmental and sustainability benefits compared to traditional chemical synthesis or extraction from natural sources. By using renewable feedstocks, microbial production processes can reduce dependency on petrochemical-derived ingredients, minimizing the environmental footprint of pharmaceutical manufacturing. Furthermore, microbial systems generally require fewer toxic reagents and generate less hazardous waste than conventional methods.

For instance, the microbial production of omega-3 fatty acids, used in supplements and cardiovascular drugs, reduces the need to harvest these compounds from fish, protecting marine ecosystems. Similarly, producing plant-based pharmaceuticals, such as paclitaxel (an anticancer drug), in microbes can reduce deforestation and habitat destruction associated with harvesting these compounds from natural sources.

9.8 CHALLENGES AND FUTURE PROSPECTS IN MICROBIAL PHARMACEUTICAL PRODUCTION

Despite the promise of microbial production systems, several challenges remain. One major hurdle is the difficulty of achieving human-like post-translational modifications in microbial systems, especially for complex proteins and antibodies. While advances in synthetic biology and glycoengineering have improved the capabilities of microbial hosts, mammalian cells are still the preferred systems for producing many biologics due to their capacity for complex protein folding and modification.

Another challenge is the regulatory landscape, as microbial production systems must meet stringent standards for pharmaceutical products. Ensuring consistency and quality in microbial production requires advanced bioprocess monitoring and control, which can increase the complexity and cost of production.

Looking forward, advances in synthetic biology, metabolic engineering, and AI are expected to further enhance microbial platforms for pharmaceutical production. By developing microorganisms capable of producing a broader range of biomolecules, including personalized medicines and biologics, microbial production will play an increasingly important role in the pharmaceutical industry. As these technologies mature, microbial systems may soon become a primary source of complex, high-value drugs, offering a sustainable and cost-effective alternative to traditional production methods.

9.9 CONCLUSION

The engineering of microorganisms for pharmaceutical production has transformed the industry, enabling the synthesis of complex biomolecules that were once difficult or costly to produce.

Through genetic engineering, synthetic biology, and metabolic pathway optimization, microbes are now capable of producing a diverse array of pharmaceuticals, from antibiotics to therapeutic proteins. As advances in AI and synthetic biology continue, microbial production systems will become even more sophisticated, efficient, and capable of meeting the growing demand for biologics and other pharmaceutical biomolecules.

By addressing current challenges and leveraging future technological innovations, microbial production has the potential to revolutionize the pharmaceutical industry, offering sustainable and scalable solutions for the manufacture of life-saving drugs.

CHAPTER 10

ENZYME AND COFACTOR ENGINEERING: CURRENT TRENDS AND FUTURE PROSPECTS IN THE PHARMACEUTICAL AND FERMENTATION INDUSTRIES

10.1 ENZYMES

Enzymes are proteins that act as biological catalysts by accelerating chemical reactions. The molecules upon which enzymes may act are called substrates, and the enzyme converts the substrates into different molecules known as products. Almost all metabolic processes in the cell need enzyme catalysis in order to occur at rates fast enough to sustain life. Metabolic pathways depend upon enzymes to catalyze individual steps. The study of enzymes is called *enzymology* and the field of pseudoenzyme analysis recognizes that during evolution, some enzymes have lost the ability to carry out biological catalysis, which is often reflected in their amino acid sequences and unusual 'pseudocatalytic' properties.

Enzymes are known to catalyze more than 5,000 biochemical reaction types. Other biocatalysts are catalytic RNA molecules, also called ribozymes. They are sometimes described as a *type* of enzyme rather than being *like* an enzyme, but even in the decades since ribozymes' discovery in 1980–1982, the word *enzyme* alone often means the protein type specifically (as is used in this article).

An enzyme's specificity comes from its unique three-dimensional structure. Like all catalysts, enzymes increase the reaction rate by lowering its activation energy. Some enzymes can make their conversion of substrate to product occur many millions of times faster. An extreme example is orotidine 5'-phosphate decarboxylase, which allows a reaction that would otherwise take millions of years to occur in milliseconds.

Chemically, enzymes are like any catalyst and are not consumed in chemical reactions, nor do they alter the equilibrium of a reaction. Enzymes differ from most other catalysts by being much more specific. Enzyme activity can be affected by other molecules: inhibitors are molecules that decrease enzyme activity, and activators are molecules that increase activity. Many therapeutic drugs and poisons are enzyme inhibitors. An enzyme's activity decreases markedly outside its optimal temperature and pH, and many enzymes are (permanently) denatured when exposed to excessive heat, losing their structure and catalytic properties.

Some enzymes are used commercially, for example, in the synthesis of antibiotics. Some household products use enzymes to speed up chemical reactions: enzymes in biological washing powders break down protein, starch or fat stains on clothes, and enzymes in meat tenderizer break down proteins into smaller molecules, making the meat easier to chew.

10.1.1 History

By the late 17th and early 18th centuries, the digestion of meat by stomach secretions and the conversion of starch to sugars by plant extracts and saliva were known but the mechanisms by which these occurred had not been identified. French chemist Anselme Payen was the first to discover an enzyme, diastase, in 1833. A few decades later, when studying the fermentation of sugar to alcohol by yeast, Louis Pasteur concluded that this fermentation was caused by a vital force contained within the yeast cells called "ferments", which were thought to function only within living organisms. He wrote that "alcoholic fermentation is an act correlated with the life and organization of the yeast cells, not with the death or putrefaction of the cells."

In 1877, German physiologist Wilhelm Kühne (1837–1900) first used the term *enzyme* to describe this process. The word *enzyme* was used later to refer to nonliving substances such as pepsin, and the word *ferment* was used to refer to chemical activity produced by living organisms. Eduard Buchner submitted his first paper on the study of yeast extracts in 1897. In a series of experiments at the University of Berlin, he found that sugar was fermented by yeast extracts even when there were no living yeast cells in the mixture. He named the enzyme that brought about the fermentation of sucrose "zymase". In 1907, he received the Nobel Prize in Chemistry for "his discovery of cell-free fermentation". Following Buchner's example, enzymes are usually named according to the reaction they carry out: the suffix *-ase* is combined with the name of the substrate (e.g., lactase is the enzyme that cleaves lactose) or to the type of reaction (e.g., DNA polymerase forms DNA polymers).

The biochemical identity of enzymes was still unknown in the early 1900s. Many scientists observed that enzymatic activity was associated with proteins, but others (such as Nobel laureate Richard Willstätter) argued that proteins were merely carriers for the true enzymes and that proteins *per se* were incapable of catalysis. In 1926, James B. Sumner showed that the enzyme urease was a pure protein and crystallized it; he did likewise for the enzyme catalase in 1937. The conclusion that pure proteins can be enzymes was definitively demonstrated by John Howard Northrop and Wendell Meredith Stanley, who worked on the digestive enzymes pepsin (1930), trypsin and chymotrypsin. These three scientists were awarded the 1946 Nobel Prize in Chemistry. The discovery that enzymes could be crystallized eventually allowed their structures to be solved by x-ray crystallography. This was first done for lysozyme, an enzyme found in tears, saliva and egg whites that digests the coating of some bacteria; the structure was solved by a group led by David Chilton Phillips and published in 1965. This high-resolution structure of lysozyme marked the beginning of the field of structural biology and the effort to understand how enzymes work at an atomic level of detail.

10.1.2 Classification and Nomenclature

Enzymes can be classified by two main criteria: either amino acid sequence similarity (and thus evolutionary relationship) or enzymatic activity.

Enzyme Activity

An enzyme's name is often derived from its substrate or the chemical reaction it catalyzes, with the word ending in *-ase*. Examples are lactase, alcohol dehydrogenase and DNA polymerase. Different enzymes that catalyze the same chemical reaction are called isozymes.

The International Union of Biochemistry and Molecular Biology have developed a nomenclature for enzymes, the EC numbers (for "Enzyme Commission"). Each enzyme is described by "EC" followed by a sequence of four numbers which represent the hierarchy of enzymatic activity (from very general to very specific). That is, the first number broadly classifies the enzyme based on its mechanism while the other digits add more and more specificity.

The top-level classification is:

- **EC 1, Oxidoreductases:** Catalyze oxidation/reduction reactions.
- **EC 2, Transferases:** Transfer a functional group (*e.g.* a methyl or phosphate group).
- **EC 3, Hydrolases:** Catalyze the hydrolysis of various bonds.
- **EC 4, Lyases:** Cleave various bonds by means other than hydrolysis and oxidation.
- **EC 5, Isomerases:** Catalyze isomerization changes within a single molecule.
- **EC 6, Ligases:** Join two molecules with covalent bonds.
- **EC 7, Translocases:** Catalyze the movement of ions or molecules across membranes, or their separation within membranes.

These sections are subdivided by other features such as the substrate, products, and chemical mechanism. An enzyme is fully specified by four numerical designations. For example, hexokinase is a transferase that adds a phosphate group to a hexose sugar, a molecule containing an alcohol group.

Sequence Similarity

EC categories do not reflect sequence similarity. For instance, two ligases of the same EC number that catalyze exactly the same reaction can have completely different sequences. Independent of their function, enzymes, like any other proteins, have been classified by their sequence similarity into numerous families. These families have been documented in dozens of different protein and protein family databases such as Pfam.

Non-Homologous Isofunctional Enzymes

Unrelated enzymes that have the same enzymatic activity have been called *non-homologous isofunctional enzymes.* Horizontal gene transfer may spread these genes to unrelated species, especially bacteria where they can replace endogenous genes of the same function, leading to hon-homologous gene displacement.

10.1.3 Structure

Enzymes are generally globular proteins, acting alone or in larger complexes. The sequence of the amino acids specifies the structure which in turn determines the catalytic activity of the enzyme. Although structure determines function, a novel enzymatic activity cannot yet be predicted from structure alone. Enzyme structures unfold (denature) when heated or exposed to chemical denaturants and this disruption to the structure typically causes a loss of activity. Enzyme denaturation is normally linked to temperatures above a species' normal level; as a result, enzymes from bacteria living in volcanic environments such as hot springs are prized by industrial users for their ability to function at high temperatures, allowing enzyme-catalysed reactions to be operated at a very high rate.

Enzymes are usually much larger than their substrates. Sizes range from just 62 amino acid residues, for the monomer of 4-oxalocrotonate tautomerase, to over 2,500 residues in the animal fatty acid synthase. Only a small portion of their structure (around 2–4 amino acids) is directly involved in catalysis: the catalytic site. This catalytic site is located next to one or more binding sites where residues orient the substrates. The catalytic site and binding site together compose the enzyme's active site. The remaining majority of the enzyme structure serves to maintain the precise orientation and dynamics of the active site.

In some enzymes, no amino acids are directly involved in catalysis; instead, the enzyme contains sites to bind and orient catalytic cofactors. Enzyme structures may also contain allosteric sites where the binding of a small molecule causes a conformational change that increases or decreases activity.

A small number of RNA-based biological catalysts called ribozymes exist, which again can act alone or in complex with proteins. The most common of these is the ribosome which is a complex of protein and catalytic RNA components.

10.1.4 Mechanism

Substrate Binding

Enzymes must bind their substrates before they can catalyse any chemical reaction. Enzymes are usually very specific as to what substrates they bind and then the chemical reaction catalysed. Specificity is achieved by binding pockets with complementary shape, charge and hydrophilic/hydrophobic characteristics to the substrates. Enzymes can therefore distinguish between very similar substrate molecules to be chemoselective, regioselective and stereospecific.

Some of the enzymes showing the highest specificity and accuracy are involved in the copying and expression of the genome. Some of these enzymes have "proof-reading" mechanisms. Here, an enzyme such as DNA polymerase catalyzes a reaction in a first step and then checks that the product is correct in a second step. This two-step process results in average error rates of less than 1 error in 100 million reactions in high-fidelity mammalian polymerases. Similar proofreading mechanisms are also found in RNA polymerase, aminoacyl tRNA synthetases and ribosomes.

Conversely, some enzymes display enzyme promiscuity, having broad specificity and acting on a range of different physiologically relevant substrates. Many enzymes possess small side activities which arose fortuitously (i.e. neutrally), which may be the starting point for the evolutionary selection of a new function.

"Lock and Key" Model

To explain the observed specificity of enzymes, in 1894 Emil Fischer proposed that both the enzyme and the substrate possess specific complementary geometric shapes that fit exactly into one another. This is often referred to as "the lock and key" model. This early model explains enzyme specificity, but fails to explain the stabilization of the transition state that enzymes achieve.

Induced Fit Model

In 1958, Daniel Koshland suggested a modification to the lock and key model: since enzymes are rather flexible structures, the active site is continuously reshaped by interactions with the substrate as the substrate interacts with the enzyme. As a result, the substrate does not simply bind to a rigid active site; the amino acid side-chains that make up the active site are molded into the precise positions that enable the enzyme to perform its catalytic function. In some cases, such as glycosidases, the substrate molecule also changes shape slightly as it enters the active site.

The active site continues to change until the substrate is completely bound, at which point the final shape and charge distribution is determined. Induced fit may enhance the fidelity of molecular recognition in the presence of competition and noise via the conformational proofreading mechanism.

Catalysis

Enzymes can accelerate reactions in several ways, all of which lower the activation energy ($\Delta G^{\ddagger}$, Gibbs free energy)

- **By Stabilizing the Transition State:**
  - Creating an environment with a charge distribution complementary to that of the transition state to lower its energy.
- **By Providing an Alternative Reaction Pathway:**
  - Temporarily reacting with the substrate, forming a covalent intermediate to provide a lower energy transition state.
- **By Destabilizing the Substrate Ground State:**
  - Distorting bound substrate(s) into their transition state form to reduce the energy required to reach the transition state.
  - By orienting the substrates into a productive arrangement to reduce the reaction entropy change (the contribution of this mechanism to catalysis is relatively small).

Enzymes may use several of these mechanisms simultaneously. For example, proteases such as trypsin perform covalent catalysis using a catalytic triad, stabilize charge build-up on the transition states using an oxyanion hole, complete hydrolysis using an oriented water substrate.

Dynamics

Enzymes are not rigid, static structures; instead they have complex internal dynamic motions – that is, movements of parts of the enzyme's structure such as individual amino acid residues, groups of residues forming a protein loop or unit of secondary structure, or even an entire protein domain.

These motions give rise to a conformational ensemble of slightly different structures that interconvert with one another at equilibrium. Different states within this ensemble may be associated with different aspects of an enzyme's function.

For example, different conformations of the enzyme dihydrofolate reductase are associated with the substrate binding, catalysis, cofactor release, and product release steps of the catalytic cycle, consistent with catalytic resonance theory.

Substrate Presentation

Substrate presentation is a process where the enzyme is sequestered away from its substrate. Enzymes can be sequestered to the plasma membrane away from a substrate in the nucleus or cytosol. Or within the membrane, an enzyme can be sequestered into lipid rafts away from its substrate in the disordered region. When the enzyme is released it mixes with its substrate. Alternatively, the enzyme can be sequestered near its substrate to activate the enzyme. For example, the enzyme can be soluble and upon activation bind to a lipid in the plasma membrane and then act upon molecules in the plasma membrane.

Allosteric Modulation

Allosteric sites are pockets on the enzyme, distinct from the active site, that bind to molecules in the cellular environment. These molecules then cause a change in the conformation or dynamics of the enzyme that is transduced to the active site and thus affects the reaction rate of the enzyme. In this way, allosteric interactions can either inhibit or activate enzymes. Allosteric interactions with metabolites upstream or downstream in an enzyme's metabolic pathway cause feedback regulation, altering the activity of the enzyme according to the flux through the rest of the pathway.

10.1.5 Cofactors

Some enzymes do not need additional components to show full activity. Others require non-protein molecules called cofactors to be bound for activity. Cofactors can be either inorganic (e.g., metal ions and iron–sulfur clusters) or organic compounds (e.g., flavin and heme). These cofactors serve many purposes; for instance, metal ions can help in stabilizing nucleophilic species within the active site. Organic cofactors can be either coenzymes, which are released from the enzyme's active site during the reaction, or prosthetic groups, which are tightly bound to an enzyme. Organic prosthetic groups can be covalently bound (e.g., biotin in enzymes such as pyruvate carboxylase).

An example of an enzyme that contains a cofactor is carbonic anhydrase, which uses a zinc cofactor bound as part of its active site. These tightly bound ions or molecules are usually found in the active site and are involved in catalysis. For example, flavin and heme cofactors are often involved in redox reactions.

Enzymes that require a cofactor but do not have one bound are called *apoenzymes* or *apoproteins*. An enzyme together with the cofactor(s) required for activity is called a *holoenzyme* (or haloenzyme). The term *holoenzyme* can also be applied to enzymes that contain multiple protein subunits, such as the DNA polymerases; here the holoenzyme is the complete complex containing all the subunits needed for activity.

Coenzymes

Coenzymes are small organic molecules that can be loosely or tightly bound to an enzyme. Coenzymes transport chemical groups from one enzyme to another. Examples include NADH, NADPH and adenosine triphosphate (ATP). Some coenzymes, such as flavin mononucleotide (FMN), flavin adenine dinucleotide (FAD), thiamine pyrophosphate (TPP), and tetrahydrofolate (THF), are derived from vitamins. These coenzymes cannot be synthesized by the body *de novo* and closely related compounds (vitamins) must be acquired from the diet. The chemical groups carried include:

- The hydride ion (H^-), carried by NAD or $NADP^+$.
- The phosphate group, carried by adenosine triphosphate.
- The acetyl group, carried by coenzyme A.
- Formyl, methenyl or methyl groups, carried by folic acid.
- The methyl group, carried by S-adenosylmethionine.

Since coenzymes are chemically changed as a consequence of enzyme action, it is useful to consider coenzymes to be a special class of substrates, or second substrates, which are common to many different enzymes. For example, about 1000 enzymes are known to use the coenzyme NADH.

Coenzymes are usually continuously regenerated and their concentrations maintained at a steady level inside the cell. For example, NADPH is regenerated through the pentose phosphate pathway and S-adenosylmethionine by methionine adenosyltransferase. This continuous regeneration means that small amounts of coenzymes can be used very intensively. For example, the human body turns over its own weight in ATP each day.

10.1.6 Thermodynamics

As with all catalysts, enzymes do not alter the position of the chemical equilibrium of the reaction. In the presence of an enzyme, the reaction runs in the same direction as it would without the enzyme, just more quickly. For example, carbonic anhydrase catalyzes its reaction in either direction depending on the concentration of its reactants:

$$CO_2 + H_2O \xrightarrow{\text{Carbonic anhydrase}} H_2CO_3 \text{ (in tissues; high } CO_2 \text{ concentration)}$$

$$CO_2 + H_2O \xleftarrow{\text{Carbonic anhydrase}} H_2CO_3 \text{ (in lungs; low } CO_2 \text{ concentration)}$$

The rate of a reaction is dependent on the activation energy needed to form the transition state which then decays into products. Enzymes increase reaction rates by lowering the energy of the transition state. First, binding forms a low energy enzyme-substrate complex (ES). Second, the enzyme stabilises the transition state such that it requires less energy to achieve compared to the uncatalyzed reaction ($ES^‡$). Finally the enzyme-product complex (EP) dissociates to release the products.

Enzymes can couple two or more reactions, so that a thermodynamically favorable reaction can be used to "drive" a thermodynamically unfavourable one so that the combined energy of the products is lower than the substrates. For example, the hydrolysis of ATP is often used to drive other chemical reactions.

10.1.7 Kinetics

Enzyme kinetics is the investigation of how enzymes bind substrates and turn them into products. The rate data used in kinetic analyses are commonly obtained from enzyme assays. In 1913 Leonor Michaelis and Maud Leonora Menten proposed a quantitative theory of enzyme kinetics, which is referred to as Michaelis–Menten kinetics. The major contribution of Michaelis and Menten was to think of enzyme reactions in two stages. In the first, the substrate binds reversibly to the enzyme, forming the enzyme-substrate complex. This is sometimes called the Michaelis–Menten complex in their honor. The enzyme then catalyzes the chemical step in the reaction and releases the product. This work was further developed by G. E. Briggs and J. B. S. Haldane, who derived kinetic equations that are still widely used today.

Enzyme rates depend on solution conditions and substrate concentration. To find the maximum speed of an enzymatic reaction, the substrate concentration is increased until a constant rate of product formation is seen. This is shown in the saturation curve on the right. Saturation happens because, as substrate concentration increases, more and more of the free enzyme is converted into the substrate-bound ES complex. At the maximum reaction rate (V_{max}) of the enzyme, all the enzyme active sites are bound to substrate, and the amount of ES complex is the same as the total amount of enzyme.

V_{max} is only one of several important kinetic parameters. The amount of substrate needed to achieve a given rate of reaction is also important. This is given by the Michaelis–Menten constant (K_m), which is the substrate concentration required for an enzyme to reach one-half its maximum reaction rate; generally, each enzyme has a characteristic K_M for a given substrate. Another useful constant is k_{cat}, also called the *turnover number*, which is the number of substrate molecules handled by one active site per second.

The efficiency of an enzyme can be expressed in terms of k_{cat}/K_m. This is also called the specificity constant and incorporates the rate constants for all steps in the reaction up to and including the first irreversible step. Because the specificity constant reflects both affinity and catalytic ability, it is useful for comparing different enzymes against each other, or the same enzyme with different substrates. The theoretical maximum for the specificity constant is called the diffusion limit and is about 10^8 to 10^9 (M^{-1} s^{-1}). At this point every collision of the enzyme with its substrate will result in catalysis, and the rate of product formation is not limited by the reaction rate but by the diffusion rate. Enzymes with this property are called *catalytically perfect* or *kinetically perfect.*

Example of such enzymes are triose-phosphate isomerase, carbonic anhydrase, acetylcholinesterase, catalase, fumarase, β-lactamase, and superoxide dismutase. The turnover of such enzymes can reach several million reactions per second.

Michaelis–Menten kinetics relies on the law of mass action, which is derived from the assumptions of free diffusion and thermodynamically driven random collision. Many biochemical or cellular processes deviate significantly from these conditions, because of macromolecular crowding and constrained molecular movement. More recent, complex extensions of the model attempt to correct for these effects.

10.1.8 Inhibition

Enzyme reaction rates can be decreased by various types of enzyme inhibitors.

Types of Inhibition

Competitive

A competitive inhibitor and substrate cannot bind to the enzyme at the same time. Often competitive inhibitors strongly resemble the real substrate of the enzyme. For example, the drug methotrexate is a competitive inhibitor of the enzyme dihydrofolate reductase, which catalyzes the reduction of dihydrofolate to tetrahydrofolate. The similarity between the structures of dihydrofolate and this drug are shown in the accompanying figure. This type of inhibition can be overcome with high substrate concentration. In some cases, the inhibitor can bind to a site other than the binding-site of the usual substrate and exert an allosteric effect to change the shape of the usual binding-site.

Non-Competitive

A non-competitive inhibitor binds to a site other than where the substrate binds. The substrate still binds with its usual affinity and hence K_m remains the same. However the inhibitor reduces the catalytic efficiency of the enzyme so that V_{max} is reduced. In contrast to competitive inhibition, non-competitive inhibition cannot be overcome with high substrate concentration.

Uncompetitive

An uncompetitive inhibitor cannot bind to the free enzyme, only to the enzyme-substrate complex; hence, these types of inhibitors are most effective at high substrate concentration. In the presence of the inhibitor, the enzyme-substrate complex is inactive. This type of inhibition is rare.

Mixed

A mixed inhibitor binds to an allosteric site and the binding of the substrate and the inhibitor affect each other. The enzyme's function is reduced but not eliminated when bound to the inhibitor. This type of inhibitor does not follow the Michaelis–Menten equation.

Irreversible

An irreversible inhibitor permanently inactivates the enzyme, usually by forming a covalent bond to the protein. Penicillin and aspirin are common drugs that act in this manner.

Functions of Inhibitors

In many organisms, inhibitors may act as part of a feedback mechanism. If an enzyme produces too much of one substance in the organism, that substance may act as an inhibitor for the enzyme at the beginning of the pathway that produces it, causing production of the substance to slow down or stop when there is sufficient amount. This is a form of negative feedback. Major metabolic pathways such as the citric acid cycle make use of this mechanism.

Since inhibitors modulate the function of enzymes they are often used as drugs. Many such drugs are reversible competitive inhibitors that resemble the enzyme's native substrate, similar to methotrexate above; other well-known examples include statins used to treat high cholesterol, and protease inhibitors used to treat retroviral infections such as HIV. A common example of an irreversible inhibitor that is used as a drug is aspirin, which inhibits the COX-1 and COX-2 enzymes that produce the inflammation messenger prostaglandin. Other enzyme inhibitors are poisons. For example, the poison cyanide is an irreversible enzyme inhibitor that combines with the copper and iron in the active site of the enzyme cytochrome c oxidase and blocks cellular respiration.

10.1.9 Biological Function

Enzymes serve a wide variety of functions inside living organisms. They are indispensable for signal transduction and cell regulation, often via kinases and phosphatases. They also generate movement, with myosin hydrolyzing adenosine triphosphate (ATP) to generate muscle contraction, and also transport cargo around the cell as part of the cytoskeleton.

Other ATPases in the cell membrane are ion pumps involved in active transport. Enzymes are also involved in more exotic functions, such as luciferase generating light in fireflies. Viruses can also contain enzymes for infecting cells, such as the

HIV integrase and reverse transcriptase, or for viral release from cells, like the influenza virus neuraminidase.

An important function of enzymes is in the digestive systems of animals. Enzymes such as amylases and proteases break down large molecules (starch or proteins, respectively) into smaller ones, so they can be absorbed by the intestines. Starch molecules, for example, are too large to be absorbed from the intestine, but enzymes hydrolyze the starch chains into smaller molecules such as maltose and eventually glucose, which can then be absorbed. Different enzymes digest different food substances. In ruminants, which have herbivorous diets, microorganisms in the gut produce another enzyme, cellulase, to break down the cellulose cell walls of plant fiber.

Metabolism

Several enzymes can work together in a specific order, creating metabolic pathways. In a metabolic pathway, one enzyme takes the product of another enzyme as a substrate. After the catalytic reaction, the product is then passed on to another enzyme. Sometimes more than one enzyme can catalyze the same reaction in parallel; this can allow more complex regulation: with, for example, a low constant activity provided by one enzyme but an inducible high activity from a second enzyme.

Enzymes determine what steps occur in these pathways. Without enzymes, metabolism would neither progress through the same steps and could not be regulated to serve the needs of the cell. Most central metabolic pathways are regulated at a few key steps, typically through enzymes whose activity involves the hydrolysis of ATP. Because this reaction releases so much energy, other reactions that are thermodynamically unfavorable can be coupled to ATP hydrolysis, driving the overall series of linked metabolic reactions.

Control of Activity

There are five main ways that enzyme activity is controlled in the cell.

Regulation

Enzymes can be either activated or inhibited by other molecules. For example, the end product(s) of a metabolic pathway are often inhibitors for one of the first enzymes of the pathway (usually the first irreversible step, called committed step), thus regulating the amount of end product made by the pathways. Such a regulatory mechanism is called a negative feedback mechanism, because the amount of the end product produced is regulated by its own concentration. Negative feedback mechanism can effectively adjust the rate of synthesis of intermediate metabolites according to the demands of the cells. This helps with effective allocations of materials and energy economy, and it prevents the excess manufacture of end products. Like other homeostatic devices, the control of enzymatic action helps to maintain a stable internal environment in living organisms.

Post-Translational Modification

Examples of post-translational modification include phosphorylation, myristoylation and glycosylation. For example, in the response to insulin, the phosphorylation of multiple enzymes, including glycogen synthase, helps control the synthesis or degradation of glycogen and allows the cell to respond to changes in blood sugar. Another example of post-translational modification is the cleavage of the polypeptide chain.

Chymotrypsin, a digestive protease, is produced in inactive form as chymotrypsinogen in the pancreas and transported in this form to the stomach where it is activated. This stops the enzyme from digesting the pancreas or other tissues before it enters the gut. This type of inactive precursor to an enzyme is known as a zymogen or proenzyme.

Quantity

Enzyme production (transcription and translation of enzyme genes) can be enhanced or diminished by a cell in response to changes in the cell's environment. This form of gene regulation is called enzyme induction.

For example, bacteria may become resistant to antibiotics such as penicillin because enzymes called beta-lactamases are induced that hydrolyse the crucial beta-lactam ring within the penicillin molecule. Another example comes from enzymes in the liver called cytochrome P450 oxidases, which are important in drug metabolism. Induction or inhibition of these enzymes can cause drug interactions. Enzyme levels can also be regulated by changing the rate of enzyme degradation. The opposite of enzyme induction is enzyme repression.

Subcellular Distribution

Enzymes can be compartmentalized, with different metabolic pathways occurring in different cellular compartments. For example, fatty acids are synthesized by one set of enzymes in the cytosol, endoplasmic reticulum and Golgi and used by a different set of enzymes as a source of energy in the mitochondrion, through β-oxidation. In addition, trafficking of the enzyme to different compartments may change the degree of protonation (e.g., the neutral cytoplasm and the acidic lysosome) or oxidative state (e.g., oxidizing periplasm or reducing cytoplasm) which in turn affects enzyme activity. In contrast to partitioning into membrane bound organelles, enzyme subcellular localisation may also be altered through polymerisation of enzymes into macromolecular cytoplasmic filaments.

Organ Specialization

In multicellular eukaryotes, cells in different organs and tissues have different patterns of gene expression and therefore have different sets of enzymes (known as isozymes) available for metabolic reactions. This provides a mechanism for regulating the overall metabolism of the organism. For example, hexokinase, the first enzyme in the glycolysis pathway, has a specialized form called glucokinase expressed in the liver and pancreas that has a lower affinity for glucose yet is more sensitive to glucose concentration. This enzyme is involved in sensing blood sugar and regulating insulin production.

Involvement in Disease

Since the tight control of enzyme activity is essential for homeostasis, any malfunction (mutation, overproduction, underproduction or deletion) of a single critical enzyme can lead to a genetic disease. The malfunction of just one type of enzyme out of the thousands of types present in the human body can be fatal. An example of a fatal genetic disease due to enzyme insufficiency is Tay–Sachs disease, in which patients lack the enzyme hexosaminidase.

One example of enzyme deficiency is the most common type of phenylketonuria. Many different single amino acid mutations in the enzyme phenylalanine hydroxylase, which catalyzes the first step in the degradation of phenylalanine, result in build-up of phenylalanine and related products. Some mutations are in the active site, directly disrupting binding and catalysis, but many are far from the active site and reduce activity by destabilising the protein structure, or affecting correct oligomerisation. This can lead to intellectual disability if the disease is untreated. Another example is pseudocholinesterase deficiency, in which the body's ability to break down choline ester drugs is impaired. Oral administration of enzymes can be used to treat some functional enzyme deficiencies, such as pancreatic insufficiency and lactose intolerance.

Another way enzyme malfunctions can cause disease comes from germline mutations in genes coding for DNA repair enzymes. Defects in these enzymes cause cancer because cells are less able to repair mutations in their genomes. This causes a slow accumulation of mutations and results in the development of cancers. An example of such a hereditary cancer syndrome is xeroderma pigmentosum, which causes the development of skin cancers in response to even minimal exposure to ultraviolet light.

10.1.10 Evolution

Similar to any other protein, enzymes change over time through mutations and sequence divergence. Given their central role in metabolism, enzyme evolution plays a critical role in adaptation. A key question is therefore whether and how enzymes can change their enzymatic activities alongside. It is generally accepted that many new enzyme activities have evolved through gene duplication and mutation of the duplicate copies although evolution can also happen without duplication. One example of an enzyme that has changed its activity is the ancestor of methionyl aminopeptidase (MAP) and creatine amidinohydrolase (creatinase) which are clearly homologous but catalyze very different reactions (MAP removes the amino-terminal methionine in new proteins while creatinase hydrolyses creatine to sarcosine and urea). In addition, MAP is metal-ion dependent while creatinase is not, hence this property was also lost over time. Small changes of enzymatic activity are extremely common among enzymes. In particular, substrate binding specificity can easily and quickly change with single amino acid changes in their substrate binding pockets. This is frequently seen in the main enzyme classes such as kinases.

10.2 THE IMPORTANCE OF ENZYME ENGINEERING IN INDUSTRY

Enzyme engineering is the process of modifying an enzyme's structure or function to enhance its performance for specific applications. In the pharmaceutical industry, enzyme engineering enables the synthesis of chiral drugs, a process that requires high stereoselectivity to ensure the production of a drug's active form. Similarly, in the fermentation industry, engineered enzymes enhance the breakdown of biomass and improve the efficiency of bioprocesses for producing biofuels, food additives, and other chemicals.

Due to their high specificity, enzymes help reduce unwanted side reactions, thereby decreasing the generation of toxic byproducts. This selectivity contributes to more sustainable processes with lower environmental impact, aligning with green chemistry principles.

Additionally, enzymes operate under milder conditions than traditional chemical catalysts, reducing the need for high temperatures, pressures, or harsh reagents. By optimizing these catalysts through engineering, industries can achieve greater productivity, efficiency, and environmental sustainability.

10.3 DIRECTED EVOLUTION: A PRIMARY TOOL FOR ENZYME ENGINEERING

Directed evolution (DE) is a method used in protein engineering that mimics the process of natural selection to steer proteins or nucleic acids toward a user-defined goal. It consists of subjecting a gene to iterative rounds of mutagenesis (creating a library of variants), selection (expressing those variants and isolating members with the desired function) and amplification (generating a template for the next round). It can be performed *in vivo* (in living organisms), or *in vitro* (in cells or free in solution).

Directed evolution is used both for protein engineering as an alternative to rationally designing modified proteins, as well as for experimental evolution studies of fundamental evolutionary principles in a controlled, laboratory environment.

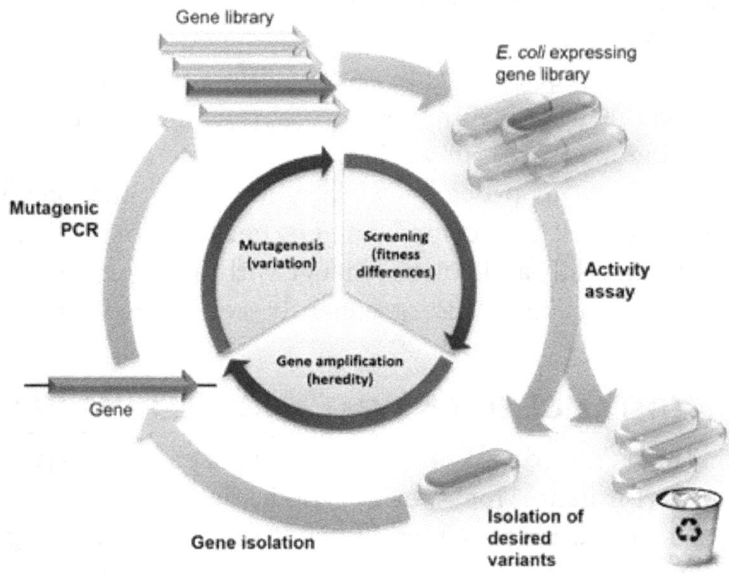

Fig 1: An Example of Directed Evolution with Comparison to Natural Evolution

The inner cycle indicates the 3 stages of the directed evolution cycle with the natural process being mimicked in brackets. The outer circle demonstrates steps in a typical experiment. The red symbols indicate functional variants, the pale symbols indicate variants with reduced function.

10.3.1 History

Directed evolution has its origins in the 1960s with the evolution of RNA molecules in the "Spiegelman's Monster" experiment. The concept was extended to protein evolution via evolution of bacteria under selection pressures that favoured the evolution of a single gene in its genome.

Early phage display techniques in the 1980s allowed targeting of mutations and selection to a single protein. This enabled selection of enhanced binding proteins, but was not yet compatible with selection for catalytic activity of enzymes. Methods to evolve enzymes were developed in the 1990s and brought the technique to a wider scientific audience.

The field rapidly expanded with new methods for making libraries of gene variants and for screening their activity. The development of directed evolution methods was honored in 2018 with the awarding of the Nobel Prize in Chemistry to Frances Arnold for evolution of enzymes, and George Smith and Gregory Winter for phage display.

10.3.2 Principles

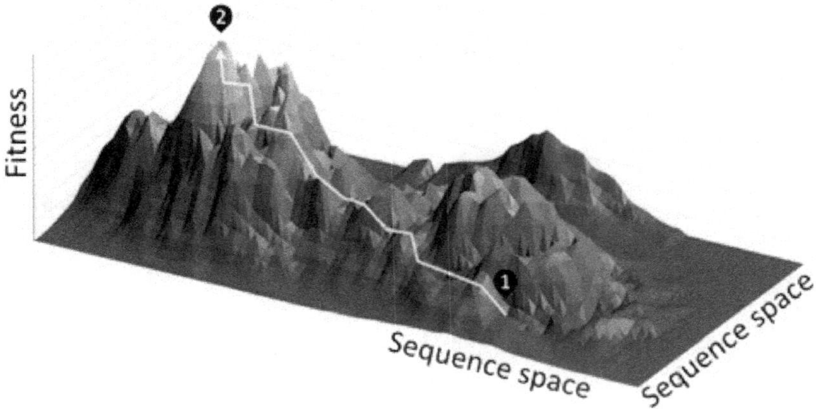

Fig 2: Directed Evolution is Analogous to Climbing a Hill on a 'Fitness Landscape' where Elevation Represents the Desired Property

Each round of selection samples mutants on all sides of the starting template and selects the mutant with the highest elevation, thereby climbing the hill. This is repeated until a local summit is reached. Directed evolution is a mimic of the natural evolution cycle in a laboratory setting. Evolution requires three things to happen: variation between replicators, that the variation causes fitness differences upon which selection acts, and that this variation is heritable. In DE, a single gene is evolved by iterative rounds of mutagenesis, selection or screening, and amplification. Rounds of these steps are typically repeated, using the best variant from one round as the template for the next to achieve stepwise improvements.

The likelihood of success in a directed evolution experiment is directly related to the total library size, as evaluating more mutants increases the chances of finding one with the desired properties.

Generating Variation

The amino acid substituted into a given position is shown. Each dot or set of connected dots is one member of the library. Error-prone PCR randomly mutates some residues to other amino acids. Alanine scanning replaces each residue of the protein with alanine, one-by-one. Site saturation substitutes each of the 20 possible amino acids (or some subset of them) at a single position, one-by-one.

The first step in performing a cycle of directed evolution is the generation of a library of variant genes. The sequence space for random sequence is vast (10^{130} possible sequences for a 100 amino acid protein) and extremely sparsely populated by functional proteins. Neither experimental, nor natural evolution can ever get close to sampling so many sequences. Of course, natural evolution samples variant sequences close to functional protein sequences and this is imitated in DE by mutagenising an already functional gene. Some calculations suggest it is entirely feasible that for all practical (i.e., functional and structural) purposes, protein sequence space has been fully explored during the course of evolution of life on Earth.

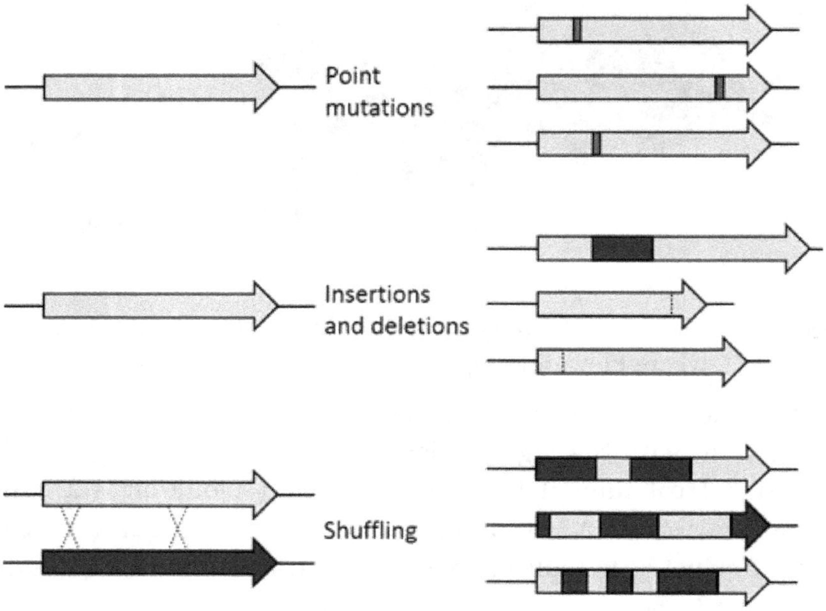

Fig 3: Starting Gene (Left) and Library of Variants (Right).

Point mutations change single nucleotides. Insertions and deletions add or remove sections of DNA. Shuffling recombines segments of two (or more) similar genes.

The starting gene can be mutagenised by random point mutations (by chemical mutagens or error prone PCR) and insertions and deletions (by transposons). Gene recombination can be mimicked by DNA shuffling of several sequences (usually of more than 70% sequence identity) to jump into regions of sequence space between the shuffled parent genes.

Finally, specific regions of a gene can be systematically randomised for a more focused approach based on structure and function knowledge. Depending on the method, the library generated will vary in the proportion of functional variants it contains. Even if an organism is used to express the gene of interest, by mutagenising only that gene the rest of the organism's genome remains the same and can be ignored for the evolution experiment (to the extent of providing a constant genetic environment).

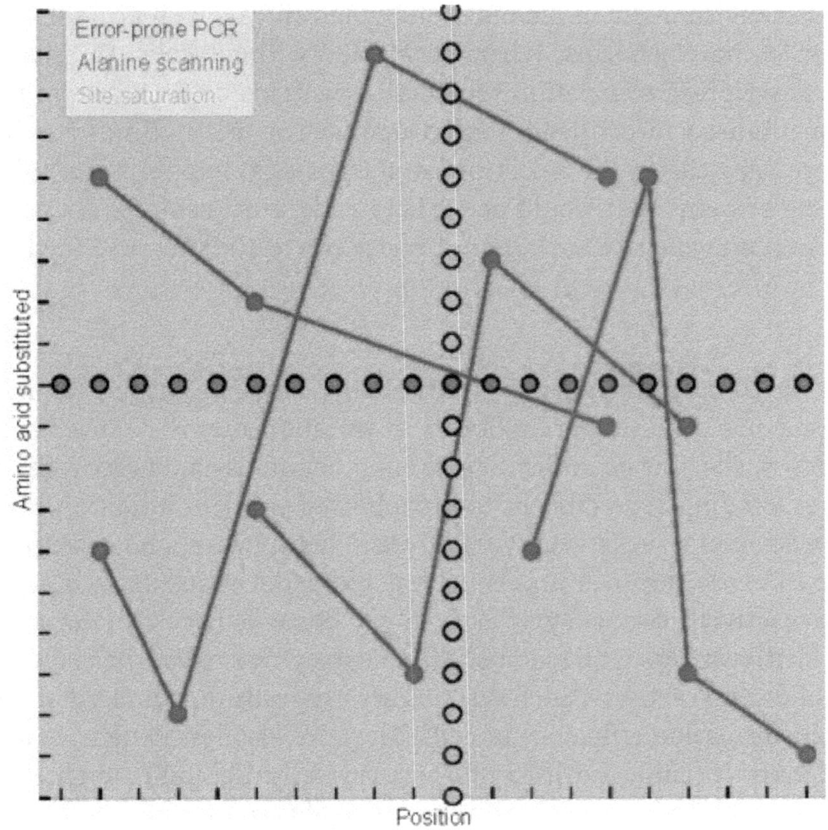

Fig 4: How DNA Libraries Generated by Random Mutagenesis Sample Sequence Space

Detecting Fitness Differences

The majority of mutations are deleterious and so libraries of mutants tend to mostly have variants with reduced activity. Therefore, a high-throughput assay is vital for measuring activity to find the rare variants with beneficial mutations that improve the desired properties. Two main categories of method exist for isolating functional variants. Selection systems directly couple protein function to survival of the gene, whereas screening systems individually assay each variant and allow a quantitative threshold to be set for sorting a variant or population of variants of a desired activity. Both selection and screening can be performed in living cells (*in vivo* evolution) or performed directly on the protein or RNA without any cells (*in vitro* evolution).

During *in vivo* evolution, each cell (usually bacteria or yeast) is transformed with a plasmid containing a different member of the variant library. In this way, only the gene of interest differs between the cells, with all other genes being kept the same. The cells express the protein either in their cytoplasm or surface where its function can be tested. This format has the advantage of selecting for properties in a cellular environment, which is useful when the evolved protein or RNA is to be used in living organisms. When performed without cells, DE involves using *in vitro* transcription translation to produce proteins or RNA free in solution or compartmentalised in artificial microdroplets. This method has the benefits of being more versatile in the selection conditions (e.g. temperature, solvent), and can express proteins that would be toxic to cells. Furthermore, *in vitro* evolution experiments can generate far larger libraries (up to 10^{15}) because the library DNA need not be inserted into cells (often a limiting step).

Selection

Selection for binding activity is conceptually simple. The target molecule is immobilised on a solid support, a library of variant proteins is flowed over it, poor binders are washed away, and the remaining bound variants recovered to isolate their genes. Binding of an enzyme to immobilised covalent inhibitor has been also used as an attempt to isolate active catalysts. This approach, however, only selects for single catalytic turnover and is not a good model of substrate binding or true substrate reactivity. If an enzyme activity can be made necessary for cell survival, either by synthesizing a vital metabolite, or destroying a toxin, then cell survival is a function of enzyme activity. Such systems are generally only limited in throughput by the transformation efficiency of cells. They are also less expensive and labour-intensive than screening, however they are typically difficult to engineer, prone to artefacts and give no information on the range of activities present in the library.

Screening

An alternative to selection is a screening system. Each variant gene is individually expressed and assayed to quantitatively measure the activity (most often by a colourgenic or fluorogenic product). The variants are then ranked and the experimenter decides which variants to use as templates for the next round of DE. Even the most high throughput assays usually have lower coverage than selection methods but give the advantage of producing detailed information on each one of the screened variants. This disaggregated data can also be used to characterise the distribution of activities in libraries which is not possible in simple selection systems. Screening systems, therefore, have advantages when it comes to experimentally characterising adaptive evolution and fitness landscapes.

Ensuring Heredity

Either way ensures that the gene can be isolated based on the activity of the encoded protein. When functional proteins have been isolated, it is necessary that their genes are too, therefore a genotype–phenotype link is required. This can be covalent, such as mRNA display where the mRNA gene is linked to the protein at the end of translation by puromycin. Alternatively the protein and its gene can be co-localised by compartmentalisation in living cells or emulsion droplets. The gene sequences isolated are then amplified by PCR or by transformed host bacteria. Either the single best sequence, or a pool of sequences can be used as the template for the next round of mutagenesis. The repeated cycles of Diversification-Selection-Amplification generate protein variants adapted to the applied selection pressures.

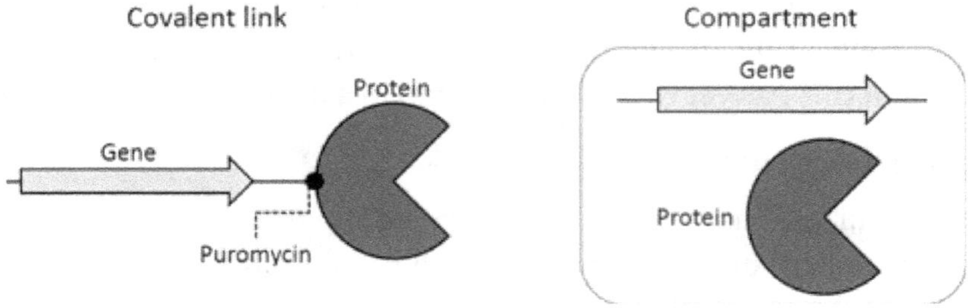

Fig 5: An Expressed Protein can Either be Covalently Linked to its Gene (as in mRNA, Left or Compartmentalized with it Cells or Artificial Compartments, Right)

10.3.3 Comparison to Rational Protein Design

Advantages of Directed Evolution

Rational design of a protein relies on an in-depth knowledge of the protein structure, as well as its catalytic mechanism. Specific changes are then made by site-directed mutagenesis in an attempt to change the function of the protein. A drawback of this is that even when the structure and mechanism of action of the protein are well known, the change due to mutation is still difficult to predict. Therefore, an advantage of DE is that there is no need to understand the mechanism of the desired activity or how mutations would affect it.

Limitations of Directed Evolution

A restriction of directed evolution is that a high-throughput assay is required in order to measure the effects of a large number of different random mutations. This can require extensive research and development before it can be used for directed evolution. Additionally, such assays are often highly specific to monitoring a particular activity and so are not transferable to new DE experiments.

Additionally, selecting for improvement in the assayed function simply generates improvements in the assayed function. To understand how these improvements are achieved, the properties of the evolving enzyme have to be measured. Improvement of the assayed activity can be due to improvements in enzyme catalytic activity or enzyme concentration. There is also no guarantee that improvement on one substrate will improve activity on another. This is particularly important when the desired activity cannot be directly screened or selected for and so a 'proxy' substrate is used. DE can lead to evolutionary specialisation to the proxy without improving the desired activity. Consequently, choosing appropriate screening or selection conditions is vital for successful DE.

The speed of evolution in an experiment also poses a limitation on the utility of directed evolution. For instance, evolution of a particular phenotype, while theoretically feasible, may occur on time-scales that are not practically feasible. Recent theoretical approaches have aimed to overcome the limitation of speed through an application of counter-diabatic driving techniques from statistical physics, though this has yet to be implemented in a directed evolution experiment.

Combinatorial Approaches

Combined, 'semi-rational' approaches are being investigated to address the limitations of both rational design and directed evolution. Beneficial mutations are rare, so large numbers of random mutants have to be screened to find improved variants. 'Focused libraries' concentrate on randomising regions thought to be richer in beneficial mutations for the mutagenesis step of DE. A focused library contains fewer variants than a traditional random mutagenesis library and so does not require such high-throughput screening.

Creating a focused library requires some knowledge of which residues in the structure to mutate. For example, knowledge of the active site of an enzyme may allow just the residues known to interact with the substrate to be randomised. Alternatively, knowledge of which protein regions are variable in nature can guide mutagenesis in just those regions.

10.3.4 Applications

Directed evolution is frequently used for protein engineering as an alternative to rational design, but can also be used to investigate fundamental questions of enzyme evolution.

Protein Engineering

As a protein engineering tool, DE has been most successful in three areas:

- Improving protein stability for biotechnological use at high temperatures or in harsh solvents.
- Improving binding affinity of therapeutic antibodies (Affinity maturation) and the activity of *de novo* designed enzymes.
- Altering substrate specificity of existing enzymes, (often for use in industry) .

Evolution Studies

The study of natural evolution is traditionally based on extant organisms and their genes. However, research is fundamentally limited by the lack of fossils (and particularly the lack of ancient DNA sequences) and incomplete knowledge of ancient environmental conditions. Directed evolution investigates evolution in a controlled system of genes for individual enzymes, ribozymes and replicators (similar to experimental evolution of eukaryotes, prokaryotes and viruses).

DE allows control of selection pressure, mutation rate and environment (both the abiotic environment such as temperature, and the biotic environment, such as other genes in the organism). Additionally, there is a complete record of all evolutionary intermediate genes.

Adaptive Laboratory Evolution of Microbial Proteomes

The natural amino acid composition of proteomes can be changed by global canonical amino acids substitutions with suitable noncanonical counterparts under the experimentally imposed selective pressure. For example, global proteome-wide substitutions of natural amino acids with fluorinated analogs have been attempted in *Escherichia coli* and *Bacillus subtilis*.

A complete tryptophan substitution with thienopyrrole-alanine in response to 20899 UGG codons in *Escherichia coli* was reported in 2015 by Budisa and Söll. The experimental evolution of microbial strains with a clear-cut accommodation of an additional amino acid is expected to be instrumental for widening the genetic code experimentally.

Directed evolution typically targets a particular gene for mutagenesis and then screens the resulting variants for a phenotype of interest, often independent of fitness effects, whereas adaptive laboratory evolution selects many genome-wide mutations that contribute to the fitness of actively growing cultures.

10.4 RATIONAL DESIGN: STRUCTURE-GUIDED ENZYME MODIFICATION

Rational design is another approach to enzyme engineering, based on a detailed understanding of an enzyme's structure and mechanism. Unlike directed evolution, which involves random mutation and screening, rational design uses computational models and structural information to predict how specific modifications will affect enzyme function.

This approach is especially valuable for optimizing catalytic efficiency, substrate specificity, and stability.

Rational design has been successfully applied to modify enzyme active sites, enhancing their affinity for industrially relevant substrates.

For example, in the pharmaceutical industry, rationally designed enzymes have been used to improve the synthesis of active pharmaceutical ingredients (APIs) by tailoring the active site for high stereoselectivity and reduced reaction time. In the fermentation industry, enzymes involved in lignocellulose breakdown, such as cellulases, have been engineered through rational design to improve substrate binding and increase catalytic efficiency, which is critical for producing bioethanol from plant biomass. One notable success is the engineering of cytochrome P450 enzymes, a class of oxidoreductases with broad applications in drug synthesis and detoxification.

By modifying key amino acids in the active site, researchers have tailored P450 enzymes to catalyze reactions that are difficult to achieve with traditional chemical catalysts, such as selective hydroxylation and epoxidation. The ability to perform these reactions under mild conditions is particularly beneficial for synthesizing complex drugs and other high-value compounds.

10.5 COFACTOR ENGINEERING: OPTIMIZING ENZYME-DEPENDENT REACTIONS

Cofactor engineering, a subset of metabolic engineering, is defined as the manipulation of the use of cofactors in an organism's metabolic pathways. In cofactor engineering, the concentrations of cofactors are changed in order to maximize or minimize metabolic fluxes.

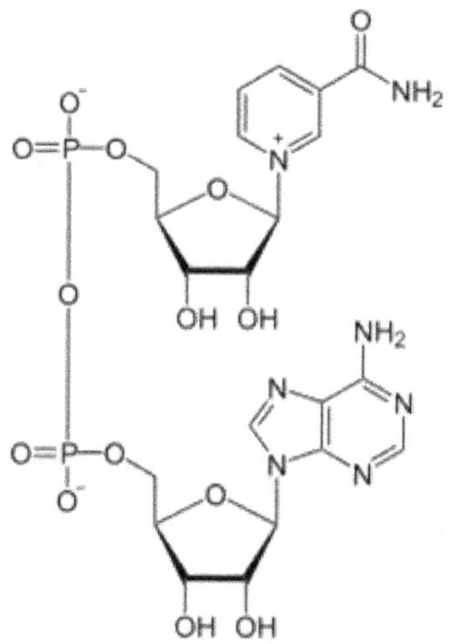

Fig 6: Common Cofactor NADH, the First Discovered

This type of engineering can be used to optimize the production of a metabolite product or to increase the efficiency of a metabolic network. The use of engineering single celled organisms to create lucrative chemicals from cheap raw materials is growing, and cofactor engineering can play a crucial role in maximizing production. The field has gained more popularity in the past decade and has several practical applications in chemical manufacturing, bioengineering and pharmaceutical industries.

Cofactors are non-protein compounds that bind to proteins and are required for the proteins normal catalytic functionality. Cofactors can be considered "helper molecules" in biological activity, and often affect the functionality of enzymes. Cofactors can be both organic and inorganic compounds.

Some examples of inorganic cofactors are iron or magnesium, and some examples of organic cofactors include ATP or coenzyme A. Organic cofactors are more specifically known as coenzymes, and many enzymes require the addition of coenzymes to assume normal catalytic function in a metabolic reaction.

The coenzymes bind to the active site of an enzyme to promote catalysis. By engineering cofactors and coenzymes, a naturally occurring metabolic reaction can be manipulated to optimize the output of a metabolic network.

10.5.1 Background

Cofactors were discovered by Arthur Harden and William Young in 1906, when they found that the rate of alcoholic fermentation in unboiled yeast extracts increased when boiled yeast extract was added. A few years after, Hans von Euler-Chelpin identified the cofactor in the boiled extract as NAD^+. Other cofactors, such as ATP and coenzyme A, were discovered later in the 1900s. The mechanism of cofactor activity was discovered when, Otto Heinrich Warburg determined in 1936 that NAD^+ functioned as an electron acceptor. Well after these initial discoveries, scientists began to realize that the manipulation of cofactor concentrations could be used as tools for the improvement of metabolic pathways.

An important group of organic cofactors is the family of molecules referred to as vitamins. Vitamin B12 (cobalamin), for example, plays a crucial role in the human body, while coenzyme B12, its derivative, is found in the metabolisms of every type of cell in our bodies. Its presence affects the synthesis and regulation of cellular DNA as well as taking part in fatty acid synthesis and energy production. Cofactors are required by many important metabolic pathways, and it is possible for the concentrations of a single type of cofactor to affect the fluxes of many different pathways . Minerals and metallic ions that organisms uptake through their diet provide prime examples of inorganic cofactors. For instance Zn^{2+} is needed to assist the enzyme carbonic anhydrase as it converts carbon dioxide and water to bicarbonate and protons. A widely recognized mineral that acts as a cofactor is iron, which is essential for the proper function of hemoglobin, the oxygen transporting protein found in red blood cells. This example in particular highlights the importance of cofactors in animal metabolism.

10.5.2 Significance

Cofactor engineering is significant in the manipulation of metabolic pathways. A metabolic pathway is a series of chemical reactions that occur in an organism.

Metabolic engineering is the subject of altering the fluxes within a metabolic pathway. In metabolic engineering, a metabolic pathway can be directly altered by changing the functionality of the enzymes involved in the pathway. Cofactor engineering, offers a distinct approach, and some advantages, to altering a metabolic pathway. Instead of changing the enzymes used in a pathway, the cofactors can be changed. This may give metabolic engineers an advantage due to certain properties of cofactors and how they can be modified.

Metabolic pathways can be used by metabolic engineers to create a desired product. By modifying the types of cofactors used and the times at which they are used, the outcome of the metabolic network can change. To create a greater production of a product, metabolic engineers have the ability to supply the network with whichever cofactor is best suited for that specific process. This leads to the optimization of networks to give a higher production of desired products. Also, changing the cofactors used in a network may be an ingenious solution to a complicated problem. A network that is present in the cell, but is often unused, may have a desirable product. Instead of engineering a completely new set of pathways to produce the product, cofactor engineering can be applied. By replacing enzymes to use cofactors readily available in a cell, the typically unused network is no longer cofactor-limited, and production may be increased.

In addition to modifying the yield of metabolic networks, changing the cofactors used in a network can reduce operation costs when trying to form a desired product. NADH and NADPH are two extremely common cellular cofactors, differing only by the presence of a phosphate group. However, this phosphate group makes NADPH much less stable than NADH, and therefore more expensive to synthesize. Thus, it is advantageous to try and use NADH is some cellular networks because it is often cheaper, more readily available, and accomplishes the same task as NADPH.

10.5.3 Tools and Processes

Cofactor engineering most often deals with the manipulation of microorganisms such as *Saccharomyces cerevisiae* and *Escherichia coli*, and as such requires the use of recombinant DNA techniques. These techniques utilize small circular segments of DNA called plasmids, which can be introduced and incorporated by microorganisms such as *Escherichia coli*. These plasmids are specifically designed in labs to be easily incorporated, and affect the expression of various protein, metabolites and enzymes. For instance, a particular plasmid may cause a change in an enzyme's amino acid sequence, which could increase its affinity for a particular substrate.

Microorganisms require a medium to grow in, and one commonly used for cultures of *Escherichia coli* is Luria-Bertani (LB) broth. This medium is often supplemented with glucose and will often contain additional molecules designed to facilitate optimal culture growth. Pre-cultures may then be grown in shake flasks. These are simply plugged Erlenmeyer flasks which are left on an orbital shaker machine, which revolves at very high RPM. This process aerates the culture, which is necessary for optimal growth. Once the pre-cultures are ready, the plasmids needed by specific experiments are added to each culture separately, and then each culture is transferred to a bio-reactor. Bio-reactors are systems which allow cultures to grow in a controlled environment. This leaves the introduced plasmids as the only independent variable. The required temperature, pH, metabolite concentrations, and various other environmental factors can be maintained by the bio-reactor ensuring identical growth conditions for each culture.

Once samples are allowed to grow in the reactor for a specified period, they can be removed and studied to determine whether the intended alterations to the organism are evident. Since cofactor engineering most often deals with metabolic pathways, these organisms are often studied but introducing specific tagged fluorescent metabolites and documenting their progression through various pathways. In other cases results are more obvious and easily observable, such as with the decreased ethanol production of yeast referred to below.

10.5.4 Applications

Changing an Enzyme's Cofactor from NADPH to NADH

Biocatalysts are required for the production of chiral building blocks needed in pharmaceuticals and other chemicals used by society. Many such biocatalysts require NADPH as a cofactor. NADPH, a cofactor quite similar to NADH, is both more expensive and less stable than its counterpart NADH. For these reasons, manufacturers would prefer that the biocatalysts they use in their production lines accept NADH over NADPH. Cofactor engineering has recently been successful in altering enzymes to prefer NADH as a cofactor instead of NADPH.

In 2010, a group of scientists performed cofactor engineering on the enzyme Gre2p, an NADPH-preferring dehydrogenase found in *Saccharomyces cerevisiae*. Gre2p reduces the compound diketone 2,5-hexanedione into the chiral building blocks (5S)-hydroxy-2-hexanone and (2S,5S)-hexanediol. The scientists determined that Asn9 (Asparagine, position 9) was an important amino acid the active site of Gre2p. Specifically, Asn9 binds to the 3'-hydroxyl group and the 2'-oxygen atom of adenyl ribose moiety.

Through direct mutagenesis, the scientists exchanged the Asn9 for both Asp (Aspartic Acid) and Glu (Glutamic Acid). This change caused Gre2p to have a decreased dependency on NADPH, and an increased affinity for NADH. This resulted in increased Gre2p activity when using NADH. It was observed that substituting

Asn9 with Glu produced a greater effect than changing Asn9 to Asp. Asn contains a polar uncharged side chain, while both Asp and Glu contain a polar charged side chain. The increased effect of Glu is caused by the extra carbon in its side chain that brings it closer to adenyl ribose moiety. This allows for stronger hydrogen bonding between the 2'- and 3'- ribose hydroxyl groups and the side chain carboxyl group. The maximum velocity of the reaction doubled, while using NADH, when Asn9 was substituted with Glu. With these results, the scientists successfully engineered Gre2p to prefer NADH over NADPH and increased the speed of 2,5-hexanedione reduction. This will allow chemical companies to decrease their manufacturing costs by using NADH instead of NADPH at least for this particular reduction.

Changing a Network's Cofactor Preference

An alternative example of changing an enzyme's preference for cofactors is to change NADH dependent reaction to NADPH dependent reactions. In this example, the enzymes themselves are not changed, but instead different enzymes are selected that accomplish the same reaction with the use of a different cofactor. An engineered pathway was created to make 1-butanol from Acetyl-CoA by changing enzymes in the metabolic pathway of S. elongatus. The Clostridium genus is known to produce 1-butanol, providing a pathway that could be inserted in S. elongatus. This pathway synthesizes 1-butanol using the reverse β-oxidation pathway. The enzymes involved in this newly engineered pathway were NADH specific, which was problematic for replicating the pathway in S. elongatus as cyanobacteria produce much more NADPH than NADH.

The research group then identified enzymes that utilize NADPH or both NADPH and NADH by bioprospecting. Acetoacetyl-CoA reductase (PhaB) was found to be a suitable replacement for hydroxybutyric dehydrogenase (Hbd). To replace AdhE2, the researchers found that NADP-dependent alcohol dehydrogenase (YqhD) from E. coli to be effective for the pathway. Furthermore, the researchers needed a dehydrogenase to replace the aldehyde dehydrogenase capacity of AdhE2. CoA-acylatingbutyraldehyde dehydrogenase (Bldh) from C. saccharoperbutylacetonicum was found to be a good suit.

Together, PhaB, Bldh, YqhD can replace Hbd and AdhE2, respectively, to change the cofactor preference of 3-ketobutyryl-CoA reduction from using NADH to using NADPH. The authors then constructed various combinations of the different enzymes (of those found in the reverse oxidation pathway and the NADPH utilizing enzymes) by overexpressing different genes in cultures of S. elongatus PCC 7942. In order to do so, they constructed plasmids containing the genes corresponding to the enzymes and combined them into the genome of S. elongates. After enzyme assays, the strain of cyanobacteria expressing the NADPH utilizing enzymes produced the greatest amount of 1-butanol (29.9 mg/L), exceeding that of strains that did not consist of the NADPH utilizing enzymes by four times. Overall, 1-butanol was produced in S. elongatus using a pathway from another organism. This pathway was modified in order to match the preferred reducing cofactor for the cyanobacteria.

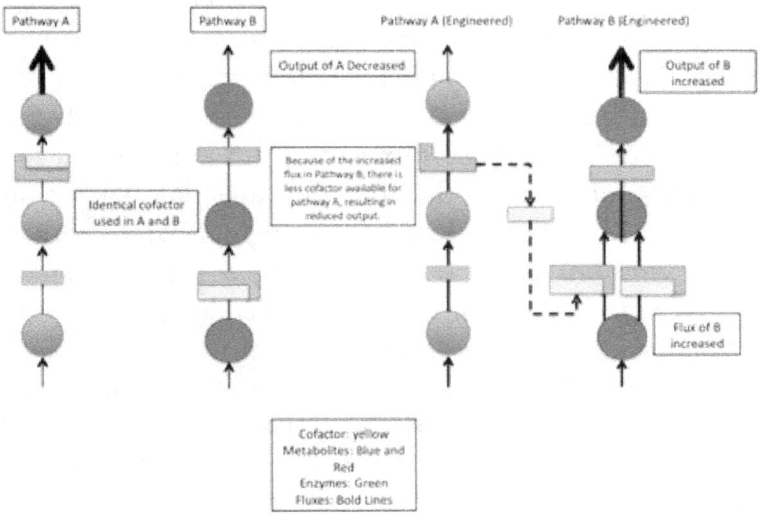

Modifying Metabolite Flux with Cofactor Equilibrium

In cofactor engineering, a metabolic pathway is altered by changing the concentrations of specific cofactors that are produced either in that particular pathway or in a separate pathway. For example, an hypothetical organism could have two arbitrary pathways called A and B where some enzymes in both A and B utilize the same cofactors. If scientists wanted to decrease the output of pathway A, they may first consider directly engineering the enzymes involved in A, perhaps to decrease a particular active site's affinity for its substrate. In some cases however, the enzymes in A may be difficult to engineer for various reasons, or it may be impossible to engineer them without dangerously affecting some third metabolic pathway C, which utilizes the same enzymes. As a separate option, scientists could increase the flux of B, which may be easier to engineer.

This in turn could "tie up" the cofactors needed by A, which would slow enzymatic activity, decreasing output in A. This is one hypothetical example of how cofactor engineering can be used, but there are many other unique cases where scientists use cofactors as a way of altering metabolic pathways.

A major advantage to cofactor engineering is that scientists can use it to successfully alter metabolic pathways that are difficult to engineer by means of ordinary metabolic engineering. This is achieved by targeting more easily engineered enzymes in separate pathways, which use the same cofactors. Since many cofactors are used by different enzymes in multiple pathways, cofactor engineering may be an efficient, cost effective alternative to current methods of metabolic engineering.

Yeast are commonly used in the beer and wine industry because they are capable of efficiently producing ethanol through the metabolic pathway fermentation in the absence of oxygen. Fermentation requires the enzyme glycerol-3-phosphate dehydrogenase (GPDH) which depends on the cofactor NADH. This pathway involves the conversion of glucose to both ethanol and glycerol, both of which use NADH as a cofactor. Scientists engineered *Saccharomyces cerevisiae* to overproduce GPDH, which shifted the cells metabolic flux away from ethanol and toward glycerol, by limiting NADH availability in the ethanol production portion of the pathway. The opposite effect was achieved by influencing a separate pathway in the cell, the Glutamate Synthesis pathway. Inactivating the expression of the enzyme glutamate dehydrogenase, which is NADPH dependent, and over expressing the enzymes glutamine synthetase and glutamate synthetase, which rely on NADH as a cofactor shifted the cofactor balance in glutamate synthesis pathway. The pathway is now dependent on NADH rather than NADPH, which decreases NADH availability in the fermentation pathway. This in turn causes increased ethanol production and decreased glycerol production. This method of manipulating metabolic fluxes could be visualized much like global fuel markets, where the increased production of ethanol for use in the automotive industry would decrease its availability in the food industry. Essentially, producing more engines which run on ethanol could result in decreased consumption of processed sweets, which contain high fructose corn syrup. This engineering of cofactors is applicable to the beer and wine industry since it allows for the regulation of ethanol levels in alcoholic beverages. Advancements in the wine industry have caused a steady increase in ethanol content, so winemakers in particular would be interested in the possibility of reducing the ethanol levels of some of their wines.

Citric Acid Cycle

Coenzyme A (CoA) and acetyl-CoA are two intermediate metabolites, most notably found in the Citric Acid Cycle, which participate in over 100 different reactions in the metabolism of microorganisms. Recent experiments have shown that over expression of the enzyme pantothenate kinase and supplementation of pantothenic acid in the CoA biosynthesis pathway have allowed adjustments of both CoA and acetyl-CoA fluxes. This increased concentration of cofactors resulted in an increased carbon flux in the isoamyl acetate synthesis pathway, increase the production efficiency of isoamyl acetate. Isoamyl acetate is used industrially for artificial flavoring and for testing the effectiveness of respirators. In addition to the production of isoamyl acetate, the manipulation of CoA biosynthesis during the pyruvate hydrogenase reaction also causes an increase in the production of both succinate and lycopene, each of which have beneficial effects on the human body. An increase in succinate concentration, which is used as a catalyst, may lead to an increase in the speed of the Citric Acid Cycle, and consequently an individual's metabolism. Increasing lycopene concentrations, has been shown to decrease the risk of prostate cancer. The potential rewards of repeating such feat of cofactor engineering and successfully incorporating them into industry practices are innumerable.

Paper Manufacturing

Many important industrial enzymes use cofactors to catalyze reactions. By using cofactors to manipulate metabolic pathways, it is possible to reduce material cost, eliminate steps in production, reduce production time, decrease pollution, and increase overall production efficiency. One case that demonstrates several of these manufacturing benefits involves the genetic engineering of aspen trees. In the paper production process, manufacturing plants must break down lignin, a biochemical compound that gives a tree trunk its stiffness, in order to form the pulp used throughout the rest of production. The chemical pulping process requires the manufacturing plant to use a significant amount of energy, as well as many expensive and toxic chemicals. A group of genetic engineers, through cofactor engineering, engineered a genetically superior aspen tree that produced less lignin. These genetically engineered trees have allowed for paper mills to reduce their costs, pollution, and manufacturing time.

10.6 INTEGRATION OF COMPUTATIONAL APPROACHES AND AI IN ENZYME ENGINEERING

The integration of computational approaches, including molecular dynamics simulations and AI-driven modeling, has transformed enzyme and cofactor engineering. Computational tools enable the prediction of enzyme structure and function, helping researchers identify regions where modifications are likely to yield beneficial effects. Molecular dynamics simulations, for instance, allow scientists to visualize the movement of enzymes and their interactions with substrates and cofactors at an atomic level. This information is invaluable for rational design and the prediction of enzyme stability and activity under different conditions.

AI and machine learning are also increasingly being used to predict enzyme activity and optimize enzyme engineering processes. Machine learning algorithms can analyze large datasets from directed evolution and rational design experiments, identifying patterns that predict which mutations will enhance enzyme performance. For example, deep learning models have been used to optimize the enzyme subtilisin, an industrial protease used in laundry detergents, increasing its stability and efficiency. By accelerating the engineering process, computational tools reduce the cost and time required to develop optimized biocatalysts, opening new possibilities for industrial applications.

10.7 FUTURE PROSPECTS: ENZYME AND PATHWAY ENGINEERING FOR METABOLIC FLEXIBILITY

As enzyme and cofactor engineering techniques continue to advance, there is a growing interest in engineering entire metabolic pathways within microorganisms, rather than focusing on single enzymes. This approach, known as pathway engineering, involves the integration of multiple engineered enzymes to optimize the flow of metabolites toward a desired product. By coordinating the activities of various enzymes and their cofactors, pathway engineering can improve yield, reduce byproducts, and enhance overall process efficiency.

One promising area for future development is the production of complex pharmaceutical compounds, such as alkaloids and terpenoids, which have valuable therapeutic properties. Engineering pathways in microbes to produce these compounds offers a sustainable alternative to extraction from plants, which is often limited by low yields and environmental concerns. For instance, engineered yeast strains have been developed to produce high levels of taxol, an anticancer compound originally sourced from the Pacific yew tree. Pathway engineering not only enables sustainable production but also provides opportunities to create novel compounds with enhanced therapeutic effects.

10.8 ENZYME AND COFACTOR ENGINEERING IN SYNTHETIC BIOLOGY

Synthetic biology combines principles of molecular biology, engineering, and computational science to create custom biological systems. Enzyme and cofactor engineering are fundamental to synthetic biology, enabling the design of enzymes and pathways for novel biochemical processes. The potential applications are vast, from producing rare pharmaceuticals to creating bio-based plastics. By engineering enzymes to function in synthetic pathways, researchers can construct microorganisms that produce new-to-nature compounds, opening up new possibilities in drug discovery and bio-based materials.

Synthetic biology also supports the design of enzyme cascades, where multiple engineered enzymes work sequentially to transform substrates into complex products. In the pharmaceutical industry, enzyme cascades have been developed for multistep drug synthesis, reducing the need for intermediate purification steps and enabling continuous biocatalysis. This approach enhances efficiency and is particularly useful for synthesizing drugs with complex structures, such as antibiotics and anti-cancer agents.

10.9 CONCLUSION

Enzyme and cofactor engineering have become indispensable tools in the pharmaceutical and fermentation industries, enabling the production of valuable biomolecules with enhanced efficiency, selectivity, and sustainability. Current trends, including directed evolution, rational design, and computational approaches, have revolutionized enzyme development, while advances in pathway and synthetic biology promise further innovation. Future prospects include the integration of AI, machine learning, and pathway engineering to create microbial factories capable of producing complex pharmaceuticals and bio-based chemicals. As these technologies continue to evolve, enzyme and cofactor engineering will play an increasingly crucial role in addressing the demands of modern industry, offering environmentally friendly, efficient, and versatile solutions for pharmaceutical and fermentation applications.

SECTION IV:

FERMENTATION MICROBIOLOGY AND BIOTECHNOLOGY: BIOCONVERSION OF RENEWABLE RESOURCES TO DESIRABLE END PRODUCTS

CHAPTER 11

CONVERSION OF RENEWABLE RESOURCES TO BIOFUELS AND FINE CHEMICALS: CURRENT TRENDS AND FUTURE PROSPECTS

11.1 UNDERSTANDING THE RENEWABLE RESOURCES

A renewable resource (also known as a flow resource) is a natural resource which will replenish to replace the portion depleted by usage and consumption, either through natural reproduction or other recurring processes in a finite amount of time in a human time scale. When the recovery rate of resources is unlikely to ever exceed a human time scale, these are called perpetual resources. Renewable resources are a part of Earth's natural environment and the largest components of its ecosphere. A positive life-cycle assessment is a key indicator of a resource's sustainability.

Definitions of renewable resources may also include agricultural production, as in agricultural products and to an extent water resources. In 1962, Paul Alfred Weiss defined renewable resources as: "*The total range of living organisms providing man with life, fibres, etc...*". Another type of renewable resources is renewable energy resources. Common sources of renewable energy include solar, geothermal and wind power, which are all categorized as renewable resources. Fresh water is an example of a renewable resource.

11.1.1 Air, Food and Water

Water Resources

Water can be considered a *renewable* material when carefully controlled usage and temperature, treatment, and release are followed. If not, it would become a non-renewable resource at that location. For example, as groundwater is usually removed from an aquifer at a rate much greater than its very slow natural recharge, it is a considered non-renewable resource. Removal of water from the pore spaces in aquifers may cause permanent compaction (subsidence) that cannot be renewed. 97.5% of the water on the Earth is salt water, and 3% is fresh water; slightly over two thirds of this is frozen in glaciers and polar ice caps. The remaining unfrozen freshwater is found mainly as groundwater, with only a small fraction (0.008%) present above ground or in the air.

Water pollution is one of the main concerns regarding water resources. It is estimated that 22% of worldwide water is used in industry. Major industrial users include hydroelectric dams, thermoelectric power plants (which use water for cooling), ore and oil refineries (which use water in chemical processes) and manufacturing plants (which use water as a solvent), it is also used for dumping garbage.

Non-Agricultural Food

Food is any substance consumed to provide nutritional support for the body. Most food has its origin in renewable resources. Food is obtained directly from plants and animals. Hunting may not be the first source of meat in the modernised world, but it is still an important and essential source for many rural and remote groups. It is also the sole source of feeding for wild carnivores.

Sustainable Agriculture

The phrase sustainable agriculture was coined by Australian agricultural scientist Gordon McClymont. It has been defined as "an integrated system of plant and animal production practices having a site-specific application that will last over the long term". Expansion of agricultural land reduces biodiversity and contributes to deforestation. The Food and Agriculture Organization of the United Nations estimates that in coming decades, cropland will continue to be lost to industrial and urban development, along with reclamation of wetlands, and conversion of forest to cultivation, resulting in the loss of biodiversity and increased soil erosion.

Although air and sunlight are available everywhere on Earth, crops also depend on soil nutrients and the availability of water. Monoculture is a method of growing only one crop at a time in a given field, which can damage land and cause it to become either unusable or suffer from reduced yields. Monoculture can also cause the build-up of pathogens and pests that target one specific species. The Great Irish Famine (1845–1849) is a well-known example of the dangers of monoculture.

Crop rotation and long-term crop rotations confer the replenishment of nitrogen through the use of green manure in sequence with cereals and other crops, and can improve soil structure and fertility by alternating deep-rooted and shallow-rooted plants. Other methods to combat lost soil nutrients are returning to natural cycles that annually flood cultivated lands (returning lost nutrients indefinitely) such as the Flooding of the Nile, the long-term use of biochar, and use of crop and livestock landraces that are adapted to less than ideal conditions such as pests, drought, or lack of nutrients.

Agricultural practices are one of the single greatest contributor to the global increase in soil erosion rates. It is estimated that "more than a thousand million tonnes of southern Africa's soil are eroded every year. Experts predict that crop yields will be halved within thirty to fifty years if erosion continues at present rates." The Dust Bowl phenomenon in the 1930s was caused by severe drought combined with farming methods that did not include crop rotation, fallow fields, cover crops, soil terracing and wind-breaking trees to prevent wind erosion.

The tillage of agricultural lands is one of the primary contributing factors to erosion, due to mechanised agricultural equipment that allows for deep plowing, which severely increases the amount of soil that is available for transport by water erosion. The phenomenon called *peak soil* describes how large-scale factory farming techniques are affecting humanity's ability to grow food in the future. Without efforts to improve soil management practices, the availability of arable soil may become increasingly problematic.

Methods to combat erosion include no-till farming, using a keyline design, growing wind breaks to hold the soil, and widespread use of compost. Fertilizers and pesticides can also have an effect of soil erosion, which can contribute to soil salinity and prevent other species from growing. Phosphate is a primary component in the chemical fertiliser applied most commonly in modern agricultural production. However, scientists estimate that rock phosphate reserves will be depleted in 50–100 years and that *Peak Phosphate* will occur in about 2030.

Industrial processing and logistics also have an effect on agriculture's sustainability. The way and locations crops are sold requires energy for transportation, as well as the energy cost for materials, labour, and transport. Food sold at a local location, such a farmers' market, have reduced energy overheads.

Air

Air is a renewable resource. All living organisms need oxygen, nitrogen (directly or indirectly), carbon (directly or indirectly) and many other gases in small quantities for their survival.

11.1.2 Non-Food Resources

An important renewable resource is wood provided by means of forestry, which has been used for construction, housing and firewood since ancient times. Plants provide the main sources for renewable resources, the main distinction is between energy crops and non-food crops. A large variety of lubricants, industrially used vegetable oils, textiles and fibre made e.g. of cotton, copra or hemp, paper derived from wood, rags or grasses, bioplastic are based on plant renewable resources. A large variety of chemical based products like latex, ethanol, resin, sugar and starch can be provided with plant renewables. Animal based renewables include fur, leather, technical fat and lubricants and further derived products, as e.g. animal glue, tendons, casings or in historical times ambra and baleen provided by whaling.

With regard to pharmacy ingredients and legal and illegal drugs, plants are important sources, however e.g. venom of snakes, frogs and insects has been a valuable renewable source of pharmacological ingredients. Before GMO production set in, insulin and important hormones were based on animal sources. Feathers, an important byproduct of poultry farming for food, is still being used as filler and as base for keratin in general. Same applies for the chitin produced in farming Crustaceans which may be used as base of chitosan. The most important part of the human body used for non-medical purposes is human hair as for artificial hair integrations, which is being traded worldwide.

Historical Role

Historically, renewable resources like firewood, latex, guano, charcoal, wood ash, plant colors as indigo, and whale products have been crucial for human needs but failed to supply demand in the beginning of the industrial era. Early modern times faced large problems with overuse of renewable resources as in deforestation, overgrazing or overfishing.

In addition to fresh meat and milk, which as food items are not the topic of this section, livestock farmers and artisans used further animal ingredients as tendons, horn, bones, bladders. Complex technical constructions as the composite bow were based on combination of animal and plant based materials. The current distribution conflict between biofuel and food production is being described as Food vs. fuel. Conflicts between food needs and usage, as supposed by fief obligations were in so far common in historical times as well. However, a significant percentage of (middle European) farmers yields went into livestock, which provides as well organic fertiliser. Oxen and horses were important for transportation purposes, drove engines as e.g. in treadmills.

Other regions solved the transportation problem with terracing, urban and garden agriculture. Further conflicts as between forestry and herding, or (sheep) herders and cattle farmers led to various solutions. Some confined wool production and sheep to large state and nobility domains or outsourced to professional shepherds with larger wandering herds.

The British Agricultural Revolution was mainly based on a new system of crop rotation, the four-field rotation. British agriculturist Charles Townshend recognised the invention in Dutch Waasland and popularised it in the 18th century UK, George Washington Carver in the USA. The system used wheat, turnips and barley and introduced as well clover. Clover is able to fix nitrogen from air, a practically non exhaustive renewable resource, into fertilizing compounds to the soil and allowed to increase yields by large. Farmers opened up a fodder crop and grazing crop. Thus livestock could to be bred year-round and winter culling was avoided. The amount of manure rose and allowed more crops but to refrain from wood pasture.

Early modern times and the 19th century saw the previous resource base partially replaced respectively supplemented by large scale chemical synthesis and by the use of fossil and mineral resources respectively. Besides the still central role of wood, there is a sort of renaissance of renewable products based on modern agriculture, genetic research and extraction technology. Besides fears about an upcoming global shortage of fossil fuels, local shortages due to boycotts, war and blockades or just transportation problems in remote regions have contributed to different methods of replacing or substituting fossil resources based on renewables.

Challenges

The use of certain basically renewable products as in TCM endangers various species. Just the black market in rhinoceros horn reduced the world's rhino population by more than 90 percent over the past 40 years.

Renewables Used for Self Sufficiency

The success of the German chemical industry till World War I was based on the replacement of colonial products. The predecessors of IG Farben dominated the world market for synthetic dyes at the beginning of the 20th century and had an important role in artificial pharmaceuticals, photographic film, agricultural chemicals and electrochemicals.

However the former Plant breeding research institutes took a different approach. After the loss of the German colonial empire, important players in the field as Erwin Baur and Konrad Meyer switched to using local crops as base for economic autarky.

Meyer as a key agricultural scientist and spatial planner of the Nazi era managed and lead Deutsche Forschungsgemeinschaft resources and focused about a third of the complete research grants in Nazi Germany on agricultural and genetic research and especially on resources needed in case of a further German war effort. A wide array of agrarian research institutes still existing today and having importance in the field was founded or enlarged in the time.

There were some major failures as trying to e.g. grow frost resistant olive species, but some success in the case of hemp, flax, rapeseed, which are still of current importance. During World War 2, German scientists tried to use Russian Taraxacum (dandelion) species to manufacture natural rubber. Rubber dandelions are still of interest, as scientists in the Fraunhofer Institute for Molecular Biology and Applied Ecology (IME) announced 2013 to have developed a cultivar that is suitable for commercial production of natural rubber.

11.1.3 Legal Situation and Subsidies

Several legal and economic means have been used to enhance the market share of renewables. The UK uses Non-Fossil Fuel Obligations (NFFO), a collection of orders requiring the electricity distribution network operators in England and Wales to purchase electricity from the nuclear power and renewable energy sectors. Similar mechanisms operate in Scotland (the Scottish Renewable Orders under the Scottish Renewables Obligation) and Northern Ireland (the Northern Ireland Non-Fossil Fuel Obligation). In the US, Renewable Energy Certificates (RECs), use a similar approach. German Energiewende is using feed-in tariffs. An unexpected outcome of the subsidies was the quick increase of pellet byfiring in conventional fossil fuel plants (compare Tilbury power stations) and cement works, making wood respectively biomass accounting for about half of Europe's renewable-energy consumption.

11.1.4 Examples of Industrial Use

Biorenewable Chemicals

Biorenewable chemicals are chemicals created by biological organisms that provide feedstocks for the chemical industry. Biorenewable chemicals can provide solar-energy-powered substitutes for the petroleum-based carbon feedstocks that currently supply the chemical industry.

The tremendous diversity of enzymes in biological organisms, and the potential for synthetic biology to alter these enzymes to create yet new chemical functionalities, can drive the chemical industry. A major platform for creation of new chemicals is the polyketide biosynthetic pathway, which generates chemicals containing repeated alkyl chain units with potential for a wide variety of functional groups at the different carbon atoms. Polyurethane research is ongoing that specifically uses renewable resources.

Bioplastics

Bioplastics are a form of plastics derived from renewable biomass sources, such as vegetable fats and oils, lignin, corn starch, pea starch or microbiota. The most common form of bioplastic is thermoplastic starch. Other forms include Cellulose bioplastics, biopolyester, Polylactic acid, and bio-derived polyethylene. The production and use of bioplastics is generally regarded as a more sustainable activity when compared to plastic production from petroleum (petroplastic); however, manufacturing of bioplastic materials is often still reliant upon petroleum as an energy and materials source. Because of the fragmentation in the market and ambiguous definitions it is difficult to describe the total market size for bioplastics, but the global production capacity is estimated at 327,000 tonnes. In contrast, global consumption of all flexible packaging is estimated at 12.3 million tonnes.

Bioasphalt

Bioasphalt is an asphalt alternative made from non-petroleum based renewable resources. Manufacturing sources of bioasphalt include sugar, molasses and rice, corn and potato starches, and vegetable oil based waste. Asphalt made with vegetable oil based binders was patented by Colas SA in France in 2004.

Renewable Energy

Renewable energy refers to the provision of energy via renewable resources which are naturally replenished as fast as they are being used. Examples are sunlight, wind, biomass, rain, tides, waves and geothermal heat. Renewable energy may replace conventional fuels in four distinct markets, namely electricity generation, hot water/space heating, motor fuels, and rural (off-grid) energy services. Manufacturing of renewable energy devices uses non-renewable resources such as mined metals and land surface.

Biomass

Biomass is referring to biological material from living, or recently living organisms, most often referring to plants or plant-derived materials. Sustainable harvesting and use of renewable resources (i.e., maintaining a positive renewal rate) can reduce air pollution, soil contamination, habitat destruction and land degradation. Biomass energy is derived from six distinct energy sources: garbage, wood, plants, waste, landfill gases, and alcohol fuels. Historically, humans have harnessed biomass-derived energy since the advent of burning wood to make fire, and wood remains the largest biomass energy source today.

However, low tech use of biomass, which still amounts for more than 10% of world energy needs may induce indoor air pollution in developing nations and results in between 1.5 million and 2 million deaths in 2000.

The biomass used for electricity generation varies by region. Forest by-products, such as wood residues, are common in the United States. Agricultural waste is common in Mauritius (sugar cane residue) and Southeast Asia (rice husks). Animal husbandry residues, such as poultry litter, are common in the UK. The biomass power generating industry in the United States, which consists of approximately 11,000 MW of summer operating capacity actively supplying power to the grid, produces about 1.4 percent of the U.S. electricity supply.

Biofuel

- A biofuel is a type of fuel whose energy is derived from biological carbon fixation. Biofuels include fuels derived from biomass conversion, as well as solid biomass, liquid fuels and various biogases.

- Bioethanol is an alcohol made by fermentation, mostly from carbohydrates produced in sugar or starch crops such as corn, sugarcane or switchgrass.
- Biodiesel is made from vegetable oils and animal fats. Biodiesel is produced from oils or fats using transesterification and is the most common biofuel in Europe.
- Biogas is methane produced by the process of anaerobic digestion of organic material by anaerobes., etc. is also a renewable source of energy.

Biogas

Biogas typically refers to a mixture of gases produced by the breakdown of organic matter in the absence of oxygen. Biogas is produced by anaerobic digestion with anaerobic bacteria or fermentation of biodegradable materials such as manure, sewage, municipal waste, green waste, plant material, and crops. It is primarily methane (CH_4) and carbon dioxide (CO_2) and may have small amounts of hydrogen sulphide (H_{2S}), moisture and siloxanes.

Natural Fibre

Natural fibres are a class of hair-like materials that are continuous filaments or are in discrete elongated pieces, similar to pieces of thread. They can be used as a component of composite materials. They can also be matted into sheets to make products such as paper or felt. Fibres are of two types: natural fibre which consists of animal and plant fibres, and man made fibre which consists of synthetic fibres and regenerated fibres.

11.1.5 Threats to Renewable Resources

Renewable resources are endangered by non-regulated industrial developments and growth. They must be carefully managed to avoid exceeding the natural world's capacity to replenish them. A life cycle assessment provides a systematic means of evaluating renewability. This is a matter of sustainability in the natural environment.

Overfishing

National Geographic has described ocean over fishing as "simply the taking of wildlife from the sea at rates too high for fished species to replace themselves." Tuna meat is driving overfishing as to endanger some species like the bluefin tuna. The European Community and other organisations are trying to regulate fishery as to protect species and to prevent their extinctions.

The United Nations Convention on the Law of the Sea treaty deals with aspects of overfishing in articles 61, 62, and 65. Examples of overfishing exist in areas such as the North Sea of Europe, the Grand Banks of North America and the East China Sea of Asia. The decline of penguin population is caused in part by overfishing, caused by human competition over the same renewable resources

Deforestation

Besides their role as a resource for fuel and building material, trees protect the environment by absorbing carbon dioxide and by creating oxygen. The destruction of rain forests is one of the critical causes of climate change. Deforestation causes carbon dioxide to linger in the atmosphere. As carbon dioxide accrues, it produces a layer in the atmosphere that traps radiation from the sun. The radiation converts to heat which causes global warming, which is better known as the greenhouse effect. Deforestation also affects the water cycle. It reduces the content of water in the soil and groundwater as well as atmospheric moisture. Deforestation reduces soil cohesion, so that erosion, flooding and landslides ensue. Rain forests house many species and organisms providing people with food and other commodities. In this way biofuels may well be unsustainable if their production contributes to deforestation.

Endangered Species

Some renewable resources, species and organisms are facing a very high risk of extinction caused by growing human population and over-consumption. It has been estimated that over 40% of all living species on Earth are at risk of going extinct. Many nations have laws to protect hunted species and to restrict the practice of hunting. Other conservation methods include restricting land development or creating preserves. The IUCN Red List of Threatened Species is the best-known worldwide conservation status listing and ranking system. Internationally, 199 countries have signed an accord agreeing to create Biodiversity Action Plans to protect endangered and other threatened species.

11.2 THE IMPORTANCE OF RENEWABLE RESOURCES

Renewable resources encompass a wide range of organic materials derived from plants, algae, and waste products. These resources include lignocellulosic biomass, agricultural residues, food waste, and dedicated energy crops. The significance of renewable resources lies not only in their abundance and availability but also in their potential to mitigate environmental impacts associated with fossil fuel use. By utilizing renewable resources, we can reduce carbon emissions, promote energy independence, and contribute to a circular economy where waste is minimized, and materials are reused.

The conversion of renewable resources into biofuels and fine chemicals aligns with the principles of sustainable development. The transition from fossil fuels to bio-based alternatives supports the achievement of global climate goals, such as those outlined in the Paris Agreement. Additionally, leveraging renewable resources can create new economic opportunities, particularly in rural areas where agricultural activities are prevalent.

11.3 CURRENT TRENDS IN BIOFUEL PRODUCTION

Biofuels, including bioethanol, biodiesel, and biogas, are produced through various biochemical and thermochemical processes that convert organic materials into usable energy sources. One of the most significant trends in biofuel production is the shift towards second-generation biofuels, which utilize non-food biomass, such as agricultural residues and energy crops, to overcome the limitations associated with first-generation biofuels that rely on food crops like corn and sugarcane.

11.3.1 Second-Generation Biofuels

Second-generation biofuels, also known as advanced biofuels, are fuels that can be manufactured from various types of non-food biomass. Biomass in this context means plant materials and animal waste used especially as a source of fuel.

First-generation biofuels are made from sugar-starch feedstocks (e.g., sugarcane and corn) and edible oil feedstocks (e.g., rapeseed and soybean oil), which are generally converted into bioethanol and biodiesel, respectively.

Second-generation biofuels are made from different feedstocks and therefore may require different technology to extract useful energy from them. Second generation feedstocks include lignocellulosic biomass or woody crops, agricultural residues or waste, as well as dedicated non-food energy crops grown on marginal land unsuitable for food production.

The term second-generation biofuels is used loosely to describe both the 'advanced' technology used to process feedstocks into biofuel, but also the use of non-food crops, biomass and wastes as feedstocks in 'standard' biofuels processing technologies if suitable. This causes some considerable confusion. Therefore it is important to distinguish between second-generation feedstocks and second-generation biofuel processing technologies.

The development of second-generation biofuels has seen a stimulus since the food vs. fuel dilemma regarding the risk of diverting farmland or crops for biofuels production to the detriment of food supply. The biofuel and food price debate involves wide-ranging views, and is a long-standing, controversial one in the literature.

Second-generation biofuel technologies have been developed to enable the use of non-food biofuel feedstocks because of concerns to food security caused by the use of food crops for the production of first-generation biofuels. The diversion of edible food biomass to the production of biofuels could theoretically result in competition with food and land uses for food crops.

First-generation bioethanol is produced by fermenting plant-derived sugars to ethanol, using a similar process to that used in beer and wine-making. This requires the use of food and fodder crops, such as sugar cane, corn, wheat, and sugar beet. The concern is that if these food crops are used for biofuel production that food prices could rise and shortages might be experienced in some countries. Corn, wheat, and sugar beet can also require high agricultural inputs in the form of fertilizers, which limit the greenhouse gas reductions that can be achieved. Biodiesel produced by transesterification from rapeseed oil, palm oil, or other plant oils is also considered a first-generation biofuel.

The goal of second-generation biofuel processes is to extend the amount of biofuel that can be produced sustainably by using biomass consisting of the residual non-food parts of current crops, such as stems, leaves and husks that are left behind once the food crop has been extracted, as well as other crops that are not used for food purposes (non-food crops), such as switchgrass, grass, jatropha, whole crop maize, miscanthus and cereals that bear little grain, and also industry waste such as woodchips, skins and pulp from fruit pressing, etc. However, its production can serve as an obstacle because it's viewed as not cost-effective as well as modern technology being insufficient for its continual creation.

The problem that second-generation biofuel processes are addressing is to extract useful feedstocks from this woody or fibrous biomass, which is predominantly composed of plant cell walls. In all vascular plants the useful sugars of the cell wall are bound within the complex carbohydrates (polymers of sugar molecules) hemicellulose and cellulose, but made inaccessible for direct use by the phenolic polymer lignin. Lignocellulosic ethanol is made by extracting sugar molecules from the carbohydrates using enzymes, steam heating, or other pre-treatments. These sugars can then be fermented to produce ethanol in the same way as first-generation bioethanol production.

The by-product of this process is lignin. Lignin can be burned as a carbon neutral fuel to produce heat and power for the processing plant and possibly for surrounding homes and businesses.

Thermochemical processes (liquefaction) in hydrothermal media can produce liquid oily products from a wide range of feedstock that has a potential to replace or augment fuels. However, these liquid products fall short of diesel or biodiesel standards. Upgrading liquefaction products through one or many physical or chemical processes may improve properties for use as fuel.

Second-Generation Technology

The following subsections describe the main second-generation routes currently under development.

Thermochemical Routes

Carbon-based materials can be heated at high temperatures in the absence (pyrolysis) or presence of oxygen, air and/or steam (gasification). These thermochemical processes yield a mixture of gases including hydrogen, carbon monoxide, carbon dioxide, methane and other hydrocarbons, and water. Pyrolysis also produces a solid char. The gas can be fermented or chemically synthesised into a range of fuels, including ethanol, synthetic diesel, synthetic gasoline or jet fuel. There are also lower temperature processes in the region of 150–374 °C, that produce sugars by decomposing the biomass in water with or without additives.

- **Gasification:** Gasification technologies are well established for conventional feedstocks such as coal and crude oil. Second-generation gasification technologies include gasification of forest and agricultural residues, waste wood, energy crops and black liquor. Output is normally syngas for further synthesis to e.g. Fischer–Tropsch products including diesel fuel, biomethanol, BioDME (dimethyl ether), gasoline via catalytic conversion of dimethyl ether, or biomethane (synthetic natural gas). Syngas can also be used in heat production and for generation of mechanical and electrical power via gas motors or gas turbines.
- **Pyrolysis:** Pyrolysis is a well established technique for decomposition of organic material at elevated temperatures in the absence of oxygen In second-generation biofuels applications forest and agricultural residues, wood waste and energy crops can be used as feedstock to produce e.g. bio-oil for fuel oil applications. Bio-oil typically requires significant additional treatment to render it suitable as a refinery feedstock to replace crude oil.
- **Torrefaction:** Torrefaction is a form of pyrolysis at temperatures typically ranging between 200–320 °C. Feedstocks and output are the same as for pyrolysis.
- **Hydrothermal Liquefaction:** Hydrothermal liquefaction is a process similar to pyrolysis that can process wet materials. The process is typically at moderate temperatures up to 400 °C and higher than atmospheric pressures. The capability to handle a wide range of materials make hydrothermal liquefaction viable for producing fuel and chemical production feedstock.

Biochemical Routes

Chemical and biological processes that are currently used in other applications are being adapted for second-generation biofuels. Biochemical processes typically employ pre-treatment to accelerate the hydrolysis process, which separates out the lignin, hemicellulose and cellulose. Once these ingredients are separated, the cellulose fractions can be fermented into alcohols.

Feedstocks are energy crops, agricultural and forest residues, food industry and municipal biowaste and other biomass containing sugars. Products include alcohols (such as ethanol and butanol) and other hydrocarbons for transportation use.

Types of Biofuel

The following second-generation biofuels are under development, although most or all of these biofuels are synthesized from intermediary products such as syngas using methods that are identical in processes involving conventional feedstocks, first-generation and second-generation biofuels. The distinguishing feature is the technology involved in producing the intermediary product, rather than the ultimate off-take. A process producing liquid fuels from gas (normally syngas) is called a gas-to-liquid (GtL) process. When biomass is the source of the gas production the process is also referred to as biomass-to-liquids (BTL).

From Syngas Using Catalysis

- Biomethanol can be used in methanol motors or blended with petrol up to 10–20% without any infrastructure changes.
- BioDME can be produced from Biomethanol using catalytic dehydration or it can be produced directly from syngas using direct DME synthesis. DME can be used in the compression ignition engine.
- Bio-derived gasoline can be produced from DME via high-pressure catalytic condensation reaction. Bio-derived gasoline is chemically indistinguishable from petroleum-derived gasoline and thus can be blended into the gasoline pool.
- Biohydrogen can be used in fuel cells to produce electricity.
- Mixed Alcohols (i.e., mixture of mostly ethanol, propanol, and butanol, with some pentanol, hexanol, heptanol, and octanol). Mixed alcohols are produced from syngas with several classes of catalysts. Some have employed catalysts similar to those used for methanol. Molybdenum sulfide catalysts were discovered at Dow Chemical and have received considerable attention. Addition of cobalt sulfide to the catalyst formulation was shown to enhance performance. Molybdenum sulfide catalysts have been well studied but have yet to find widespread use. These catalysts have been a focus of efforts at the U.S. Department of Energy's Biomass Program in the Thermochemical Platform. Noble metal catalysts have also been shown to produce mixed alcohols. Most R&D in this area is concentrated in producing mostly ethanol. However, some

fuels are marketed as mixed alcohols. Mixed alcohols are superior to pure methanol or ethanol, in that the higher alcohols have higher energy content. Also, when blending, the higher alcohols increase compatibility of gasoline and ethanol, which increases water tolerance and decreases evaporative emissions. In addition, higher alcohols have also lower heat of vaporization than ethanol, which is important for cold starts.

- Biomethane (or Bio-SNG) via the Sabatier reaction

From Syngas Using Fischer–Tropsch

The Fischer–Tropsch (FT) process is a gas-to-liquid (GtL) process. When biomass is the source of the gas production the process is also referred to as biomass-to-liquids (BTL). A disadvantage of this process is the high energy investment for the FT synthesis and consequently, the process is not yet economic.

- FT diesel can be mixed with fossil diesel at any percentage without need for infrastructure change and moreover, synthetic kerosene can be produced.

Biocatalysis

- Biohydrogen might be accomplished with some organisms that produce hydrogen directly under certain conditions. Biohydrogen can be used in fuel cells to produce electricity.
- Butanol and Isobutanol via recombinant pathways expressed in hosts such as E. coli and yeast, butanol and isobutanol may be significant products of fermentation using glucose as a carbon and energy source.
- DMF (2,5-Dimethylfuran). Recent advances in producing DMF from fructose and glucose using catalytic biomass-to-liquid process have increased its attractiveness.

Other Processes

- **HTU (Hydro Thermal Upgrading) Diesel** is produced from wet biomass. It can be mixed with fossil diesel in any percentage without need for infrastructure.
- **Wood Diesel:** A new biofuel was developed by the University of Georgia from woodchips. The oil is extracted and then added to unmodified diesel engines. Either new plants are used or planted to replace the old plants. The charcoal byproduct is put back into the soil as a fertilizer. According to the director Tom Adams since carbon is put back into the soil, this biofuel can actually be carbon negative not just carbon neutral. Carbon negative decreases carbon dioxide in the air reversing the greenhouse effect not just reducing it.

Second Generation Feedstocks

To qualify as a second generation feedstock, a source must not be suitable for human consumption. Second-generation biofuel feedstocks include specifically grown inedible energy crops, cultivated inedible oils, agricultural and municipal wastes, waste oils, and algae. Nevertheless, cereal and sugar crops are also used as feedstocks to second-generation processing technologies. Land use, existing biomass industries and relevant conversion technologies must be considered when evaluating suitability of developing biomass as feedstock for energy.

Energy Crops

Plants are made from lignin, hemicellulose and cellulose; second-generation technology uses one, two or all of these components. Common lignocellulosic energy crops include wheat straw, *Arundo donax*, *Miscanthus* spp., short rotation coppice poplar and willow. However, each offers different opportunities and no one crop can be considered 'best' or 'worst'.

Municipal Solid Waste

Municipal Solid Waste comprises a very large range of materials, and total waste arisings are increasing. In the UK, recycling initiatives decrease the proportion of waste going straight for disposal, and the level of recycling is increasing each year. However, there remains significant opportunities to convert this waste to fuel via gasification or pyrolysis.

Green Waste

Green waste such as forest residues or garden or park waste may be used to produce biofuel via different routes. Examples include Biogas captured from biodegradable green waste, and gasification or hydrolysis to syngas for further processing to biofuels via catalytic processes.

Black Liquor

Black liquor, the spent cooking liquor from the kraft process that contains concentrated lignin and hemicellulose, may be gasified with very high conversion efficiency and greenhouse gas reduction potential to produce syngas for further synthesis to e.g. biomethanol or BioDME. The yield of crude tall oil from process is in the range of 30 – 50 kg / ton pulp.

Greenhouse Gas Emissions

Lignocellulosic biofuels reduces greenhouse gas emissions by 60–90% when compared with fossil petroleum, which is on par with the better of current biofuels of the first-generation, where typical best values currently is 60–80%. In 2010, average savings of biofuels used within EU was 60%. In 2013, 70% of the biofuels used in Sweden reduced emissions with 66% or higher.

Commercial Development

An operating lignocellulosic ethanol production plant is located in Canada, run by Iogen Corporation. The demonstration-scale plant produces around 700,000 litres of bioethanol each year. A commercial plant is under construction. Many further lignocellulosic ethanol plants have been proposed in North America and around the world.

The Swedish specialty cellulose mill Domsjö Fabriker in Örnsköldsvik, Sweden develops a biorefinery using Chemrec's black liquor gasification technology. When commissioned in 2015 the biorefinery will produce 140,000 tons of biomethanol or 100,000 tons of BioDME per year, replacing 2% of Sweden's imports of diesel fuel for transportation purposes. In May 2012 it was revealed that Domsjö pulled out of the project, effectively killing the effort.

In the UK, companies like INEOS Bio and British Airways are developing advanced biofuel refineries, which are due to be built by 2013 and 2014 respectively. Under favourable economic conditions and strong improvements in policy support, NNFCC projections suggest advanced biofuels could meet up to 4.3 per cent of the UK's transport fuel by 2020 and save 3.2 million tonnes of CO_2 each year, equivalent to taking nearly a million cars off the road.

Helsinki, Finland, 1 February 2012 – UPM is to invest in a biorefinery producing biofuels from crude tall oil in Lappeenranta, Finland. The industrial scale investment is the first of its kind globally.

The biorefinery will produce annually approximately 100,000 tonnes of advanced second-generation biodiesel for transport. Construction of the biorefinery will begin in the summer of 2012 at UPM's Kaukas mill site and be completed in 2014. UPM's total investment will amount to approximately EUR 150 million.

Calgary, Alberta, 30 April 2012 – Iogen Energy Corporation has agreed to a new plan with its joint owners Royal Dutch Shell and Iogen Corporation to refocus its strategy and activities.

Shell continues to explore multiple pathways to find a commercial solution for the production of advanced biofuels on an industrial scale, but the company will NOT pursue the project it has had under development to build a larger scale cellulosic ethanol facility in southern Manitoba.

In India, Indian Oil Companies have agreed to build seven second generation refineries across the country. The companies who will be participating in building of 2G biofuel plants are Indian Oil Corporation (IOCL), HPCL and BPCL. In May 2018, the Government of India unveiled a biofuel policy wherein a sum of INR 5,000 crores was allocated to set-up 2G biorefineries. Indian oil marketing companies were in a process of constructing 12 refineries with a capex of INR 10,000 crores.

11.3.2 Algal Biofuels

Another exciting trend in biofuel production is the use of microalgae and macroalgae as feedstocks. Algal biofuels have garnered attention due to their high lipid content and rapid growth rates. Microalgae can be cultivated in a variety of environments, including freshwater, saltwater, and wastewater, making them a versatile resource for biofuel production. The extraction of lipids from microalgae can yield biodiesel, while the remaining biomass can be processed into biochar, animal feed, or fine chemicals.

Algal biofuels offer several advantages over traditional biofuels, including a higher energy density and the ability to utilize non-arable land. However, challenges remain in terms of cultivation, harvesting, and processing efficiency. Ongoing research focuses on optimizing algal cultivation systems and developing cost-effective extraction methods to make algal biofuels commercially viable.

11.4 FINE CHEMICALS FROM RENEWABLE RESOURCES

In addition to biofuels, renewable resources can be converted into fine chemicals, which are specialized chemical products used in pharmaceuticals, agrochemicals, and other high-value applications. Fine chemicals often require precise synthesis methods and high purity levels, making the use of renewable resources an attractive option for sustainable production.

11.4.1 Biocatalysis in Fine Chemical Production

Biocatalysis, which utilizes enzymes or whole cells as catalysts for chemical reactions, has gained traction in the production of fine chemicals from renewable resources. Enzymes offer several advantages, including high selectivity, mild reaction conditions, and the ability to catalyze complex reactions. By engineering enzymes and optimizing reaction conditions, researchers can develop efficient biocatalytic processes for the synthesis of various fine chemicals.

For example, enzymes can be used to produce chiral intermediates for pharmaceuticals, reducing the need for extensive purification processes. Additionally, biocatalysis can enable the conversion of renewable feedstocks into high-value chemicals, such as bioplastics and bio-based surfactants. The development of enzyme libraries and high-throughput screening techniques has accelerated the discovery of novel biocatalysts with desired properties, facilitating the expansion of renewable resource-based fine chemical production.

11.4.2 Integrated Biorefineries

Integrated biorefineries represent a holistic approach to converting renewable resources into biofuels and fine chemicals. These facilities utilize a variety of biomass feedstocks and employ multiple conversion technologies to maximize resource utilization and minimize waste. By integrating different processes, such as fermentation, thermochemical conversion, and biocatalysis, biorefineries can produce a diverse range of products while optimizing energy and resource efficiency.

The concept of biorefineries aligns with the principles of circular economy, where waste from one process can serve as a feedstock for another. For instance, lignocellulosic residues generated during biofuel production can be processed into fine chemicals or used to generate heat and power for the biorefinery. This integrated approach not only enhances economic viability but also reduces the environmental impact of biomass conversion.

11.5 FUTURE PROSPECTS: INNOVATIONS AND CHALLENGES

As the demand for biofuels and fine chemicals continues to grow, several future prospects and challenges arise in the conversion of renewable resources. Advances in technology, research, and policy frameworks will play a crucial role in shaping the future of this sector.

11.5.1 Genetic Engineering and Synthetic Biology

The application of genetic engineering and synthetic biology holds great potential for improving the efficiency of renewable resource conversion. By modifying microbial strains and plants, researchers can enhance biomass yield, optimize metabolic pathways, and improve the production of biofuels and fine chemicals. For example, synthetic biology approaches can enable the engineering of microbial communities that can effectively degrade complex lignocellulosic materials and convert them into desired products.

Furthermore, metabolic engineering can be applied to enhance the production of valuable metabolites from algal biomass. By optimizing metabolic pathways, researchers can increase the lipid content of microalgae or engineer them to produce high-value chemicals, such as pharmaceuticals or nutraceuticals. The combination of genetic engineering and synthetic biology will pave the way for more efficient and sustainable conversion processes in the future.

11.5.2 Life Cycle Assessment and Sustainability

As the biofuel and fine chemical industries expand, the importance of conducting life cycle assessments (LCAs) becomes increasingly evident. LCAs evaluate the environmental impacts associated with the entire production process, from raw material extraction to end-of-life disposal. By assessing the sustainability of different production pathways, stakeholders can make informed decisions regarding feedstock selection, processing technologies, and overall resource utilization.

The implementation of sustainable practices, such as using waste materials as feedstocks and minimizing energy consumption, will be essential for ensuring the long-term viability of biofuel and fine chemical production. Policymakers, industry leaders, and researchers must collaborate to establish guidelines and standards that promote sustainable practices throughout the supply chain.

11.5.3 Economic Viability and Market Adoption

The economic viability of biofuels and fine chemicals from renewable resources is a critical factor in their widespread adoption. While technological advancements have improved production efficiency, the cost of feedstocks, processing, and infrastructure development remains a challenge. Research and investment are needed to reduce production costs and improve the competitiveness of bio-based products in the market.

Moreover, government policies and incentives play a vital role in fostering the growth of the bioeconomy. Support for research and development, subsidies for biofuel production, and the establishment of renewable fuel standards can create a conducive environment for the commercialization of bio-based products. As public awareness of environmental issues continues to rise, market demand for sustainable alternatives will likely increase, further driving investment in renewable resource conversion.

11.6 CONCLUSION

The conversion of renewable resources into biofuels and fine chemicals represents a pivotal opportunity to address global energy and environmental challenges. Current trends, including the development of second-generation biofuels, algal biofuels, and biocatalysis, highlight the potential of renewable resources in creating sustainable energy solutions and high-value products. Looking ahead, advancements in genetic engineering, synthetic biology, and life cycle assessment will be crucial in enhancing the efficiency and sustainability of these processes. By fostering innovation and collaboration among stakeholders, the bioeconomy can contribute to a more sustainable future, reducing reliance on fossil fuels and promoting a circular economy that values renewable resources.

CHAPTER 12

BIOREFINERIES: INDUSTRIAL PERSPECTIVES AND CHALLENGES IN PRIMARY AND SECONDARY METABOLISM

Biorefineries are innovative facilities that convert renewable biomass into a spectrum of bio-based products, including biofuels, chemicals, and materials. They represent a significant shift towards sustainable production methods, aiming to reduce dependence on fossil fuels and enhance environmental stewardship. As the global demand for energy and materials continues to rise, biorefineries offer a promising pathway to integrate various biological processes for efficient resource utilization. However, they also face challenges related to primary and secondary metabolism that must be addressed to realize their full potential.

12.1 UNDERSTANDING BIOREFINERIES

A biorefinery is a refinery that converts biomass to energy and other beneficial byproducts (such as chemicals). The International Energy Agency Bioenergy Task 42 defined biorefining as "the sustainable processing of biomass into a spectrum of bio-based products (food, feed, chemicals, materials) and bioenergy (biofuels, power and/or heat)". As refineries, biorefineries can provide multiple chemicals by fractioning an initial raw material (biomass) into multiple intermediates (carbohydrates, proteins, triglycerides) that can be further converted into value-added products. Each refining phase is also referred to as a "cascading phase". The use of biomass as feedstock can provide a benefit by reducing the impacts on the environment, as lower pollutants emissions and reduction in the emissions of hazard products. In addition, biorefineries are intended to achieve the following goals:

- Supply the current fuels and chemical building blocks.
- Supply new building blocks for the production of novel materials with disruptive characteristics.
- Creation of new jobs, including rural areas.
- Valorization of waste (agricultural, urban, and industrial waste).
- Achieve the ultimate goal of reducing GHG emissions.

12.1.1 Classification of Biorefinery Systems

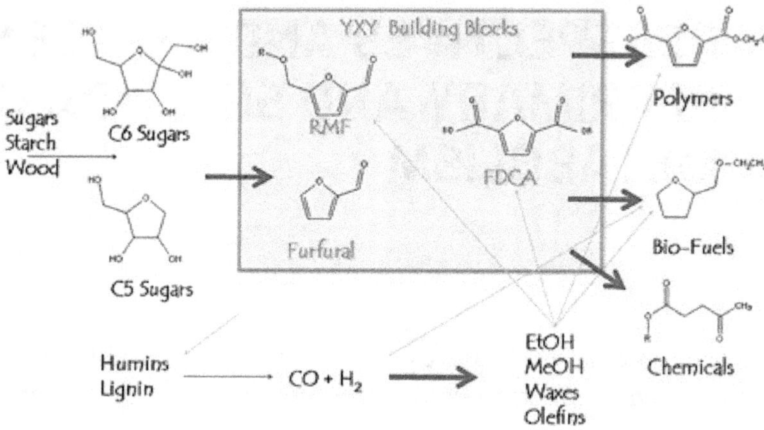

Fig 1: Chemical Diagram of the Activity of a Biorefinery

Biorefineries can be classified based in four main features:

- **Platforms:** Refers to key intermediates between raw material and final products. The most important intermediates are:
  - Biogas from anaerobic digestion.
  - Syngas from gasification.
  - Hydrogen from water-gas shift reaction, steam reforming, water electrolysis and fermentation.
  - C6 sugars from hydrolysis of sucrose, starch, cellulose and hemicellulose.
  - C5 sugars (e.g., xylose, arabinose: $C_5H_{10}O_5$), from hydrolysis of hemicellulose and food and feed side streams.
  - Lignin from the processing of lignocellulosic biomass.
  - Liquid from pyrolysis (pyrolysis oil).
- **Products:** Biorefineries can be grouped in two main categories according to the conversion of biomass in an energetic or non-energetic product. In this classification the main market must be identified:
  - *Energy-Driven Biorefinery Systems:* The main product is a second energy carrier as biofuels, power and heat.
  - *Material-Driven Biorefinery Systems:* The main product is a biobased product
- **Feedstock:** Dedicated feedstocks (Sugar crops, starch crops, lignocellulosic crops, oil-based crops, grasses, marine biomass); and

residues (oil-based residues, lignocellulosic residues, organic residues and others)

- **Processes:** Conversion process to transform biomass into a final product:
  - *Mechanical/Physical:* The chemical structure of the biomass components is preserved. This operation includes pressing, milling, separation, distillation, among others
  - *Biochemical:* Processes under low temperature and pressure, using micro-organism or enzymes.
  - *Chemical Processes:* The substrate suffer change by the action of an external chemical (e.g., hydrolysis, transesterification, hydrogenation, oxidation, pulping)
  - *Thermochemical:* Severe conditions are apply to the feedstock (high pressure and high temperature, with or without catalyst).

The aforementioned features are used to classified biorefineries systems according to the following method:

- Identify the feedstock, the main technologies included in the process, platform, and the final products.
- Draw the scheme of the refinery using the features identified in step 1.
- Label the refinery system according by citing the number of platforms, products, feedstock, and processes involved.
- Elaborate a table with the features identified, and the source of internal energy demand.

Some examples of classifications are:

- C6 sugar platform biorefinery for bioethanol and animal feed from starch crops.
- Syngas platform biorefinery for FT-diesel and phenols from straw.
- C6 and C5 sugar and syngas platform biorefinery for bioethanol, FT-diesel and furfural from saw mill residues.

12.1.2 Economic Viability of Biorefinery Systems

Techno-economic assessment (TEA) is a methodology to evaluate whether a technology or process is economically attractive. TEA research has been developed to provide information about the performance of the biorefinery concept in diverse production systems as sugarcane mills, biodiesel production, pulp and paper mills, and the treatment of industrial and municipal solid waste. Bioethanol plants and sugarcane mills are well-established processes where the biorefinery concept can be implemented since sugarcane bagasse is a feasible feedstock to produce fuels and chemicals; lignocellulosic bioethanol (2G) is produced in Brazil in two plants with capacities of 40 and 84 Ml/y (about 0.4% of the production capacity in Brazil). TEA of ethanol production using mild liquefaction of bagasse plus simultaneous saccharification and co-fermentation shows a minimum selling price between 50.38 and 62.72 US cents/L which is comparable with the market price.

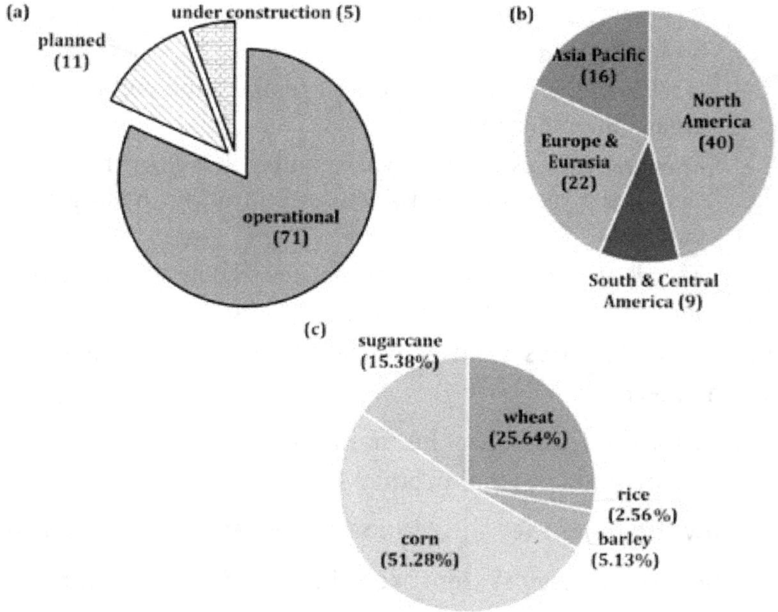

Fig 2: (a) Counts of Operational, Planned and Under-Construction Cellulosic Biorefineries with Biochemical Conversion Technology, (b) Global Distribution of Plants and (c) Shares of Corn, Wheat, Rice, Barley and Sugarcane

The production of xylitol, citric acid and glutamic acid from sugarcane lignocellulose (bagasse and harvesting residues), each in combination with electricity have been evaluated; the three biorefinery systems were simulated to be annexed to an existing sugar mill in South Africa. The production of xylitol and glutamic acid has shown economic feasibility with an Internal Rate of Return (IRR) of 12.3% and 31.5%, exceeding the IRR of the base case (10.3%). Likewise, the production of ethanol, lactic acid or methanol and ethanol-lactic acid from sugarcane bagasse have been studied; lactic acid demonstrated to be economically attractive by showing the greatest net present value (M$476–1278); in the same way; the production of ethanol and lactic acid as co-product was found to be a favorable scenario (net present value between M$165 and M$718) since this acid has applications in the pharmaceutical, cosmetic, chemical and food industry.

As for biodiesel production, this industry also has the potential to integrate biorefinery systems to convert residual biomasses and wastes into biofuel, heat, electricity and bio-based green products. Glycerol is the main co-product in biodiesel production and can be transformed into valuable products through chemocatalytic technologies; the valorization of glycerol for the production of lactic acid, acrylic acid, allyl alcohol, propanediols, and glycerol carbonate has been evaluated; all glycerol valorization routes shown to be profitable, being the most attractive the manufacture of glycerol carbonate.

Palm empty fruit bunches (EFB) are an abundant lignocellulosic residues from the palm oil/biodiesel industry, the conversion of this residue into ethanol, heat and power, and cattle feed were evaluated according to techno-economic principles, the scenarios under study shown reduced economic benefits, although their implementation represented a reduction in the environmental impact (climate change and fossil fuel depletion) compared to the traditional biodiesel production. The economic feasibility for bio-oil production from EFB via fast pyrolysis using the fluidized-bed was studied, crude bio-oil can potentially be produced from EFB at a product value of 0.47 $/kg with a payback period and return on investment of 3.2 years and 21.9%, respectively.

The integration of microalgae and *Jatropha* as a viable route for the production of biofuels and biochemicals has been analyzed in the United Arab Emirates (UAE) context. Three scenarios were examined; in all of them, biodiesel and glycerol is produced; in the first scenario biogas and organic fertilizer is produced by anaerobic fermentation of *Jatropha* fruit cake and seedcake; the second scenario includes the production of lipids from *Jatropha* and microalgae to produce biodiesel and the production of animal feed, biogas and organic fertilizer; the third scenario involves the production of lipids from microalgae for the production of biodiesel as well as hydrogen and animal feed as final product; only the first scenario was profitable.

In regard to the pulp and paper industry; lignin is a natural polymer co-generated and is generally used as boiler fuel to generate heat or steam to cover the energy demand in the process. Since lignin accounts for 10–30 wt% of the available lignocellulosic biomass and is equivalent to ~40% of its energy contents; the economics of biorefineries depend on the cost-effective processes to transform lignin into value-added fuels and chemicals.

The conversion of an existing Swedish kraft pulp mill to the production of dissolving pulp, electricity, lignin, and hemicellulose has been studied; self-sufficiency in terms of steam and the production of excess steam was a key factor for the integration of a lignin separation plant; in this case; the digester has to be upgraded for preserving the same production level and represents 70% of the total investment cost of conversion. The potential of using the kraft process for producing bioethanol from softwoods in a repurposed or co-located kraft mill has been studied, a sugar recovery higher than 60% enables the process to be competitive for the production of ethanol from softwood.

The repurposing of a kraft pulp mill to produce both ethanol and dimethyl ether has been investigated; in the process, cellulose is separated by and an alkaline pretreatment and then is hydrolyzed and fermented to produce ethanol, while the resulting liquor containing dissolved lignin is gasified and refined to dimethyl ether; the process demonstrate to be self-sufficient in terms of hot utility (fresh steam) demand but with a deficit of electricity; the process can be feasible, economically speaking, but is highly dependent on the development of biofuel prices. The exergetic and economic evaluation for the production of catechol from lignin was performed to determine its feasibility; the results showed that the total capital investment was 4.9 M$ based on the plant capacity of 2,544 kg/d of feedstock; besides, the catechol price was estimated to be 1,100 $/t and the valorization ratio was found to be 3.02.

The high generation of waste biomass is an attractive source for conversion to valuable products, several biorefinery routes has been proposed to upgrade waste streams in valuable products. The production of biogas from banana peel (*Musa x paradisiaca*) under the biorefinery concept is a promissory alternative since is possible to obtain biogas and other co-products including ethanol, xylitol, syngas, and electricity; this process also provides high profitability for high production scales. The economic assessment of the integration of organic waste anaerobic digestion with other mixed culture anaerobic fermentation technologies was studied; the highest profit is obtained by dark fermentation of food waste with separation and purification of acetic and butyric acids (47 USD/t of food waste). The technical feasibility, profitability and extent of investment risk to produce sugar syrups from food and beverage waste was analyzed; the returns on investment shown to be satisfactory for the production of fructose syrup (9.4%), HFS42 (22.8%) and glucose-rich syrup (58.9%); the sugar syrups also have high cost competitiveness with relatively low net production costs and minimum selling prices. The valorization of municipal solid waste through integrated mechanical biological chemical treatment (MBCT) systems for the production of levulinic acid has been studied, the revenue from resource recovery and product generation (without the inclusion of gate fees) is more than enough to out- weigh the waste collection fees, annual capital and operating costs.

12.1.3 Environmental Impact of Biorefinery Systems

One of the main goals of biorefineries is to contribute to a more sustainable industry by the conservation of resources and by reducing greenhouse gas emissions and other pollutants. Nevertheless, other environmental impacts may be associated to the production of biobased products; as land use change, eutrophication of water, the pollution of the environment with pesticides, or higher energy and material demand that lead to environmental burdens. Life cycle assessment

(LCA) is a methodology to evaluate the environmental load of a process, from the extraction of raw materials to the end use. LCA can be used to investigate the potential benefits of biorefinery systems; multiple LCA studies has been developed to analyse whether biorefineries are more environmentally friendly compared to conventional alternatives.

Feedstock is one of the main sources of environmental impacts in the biofuel production, the source of this impacts are related to the field operation to grow, handle and transport the biomass to the biorefinery gate. Agricultural residues are the feedstock with the lowest environmental impact followed by lignocellulosic crops; and finally by first-generation arable crops, although the environmental impacts are sensitive to factors such as crop management practices, harvesting systems, and crop yields.

The production of chemicals from biomass feedstock has shown environmental benefits; bulk chemicals from biomass-derived feedstocks have been studied showing savings on non renewable energy use and greenhouse gas emissions.

The environmental assessment for 1G and 2G ethanol shows that these two biorefinery systems are able to mitigate climate change impacts in comparison to gasoline, but higher climate change benefits are achieved with 2G ethanol production (up to 80% reduction).

The conversion of palm empty fruit bunches into valuable products (ethanol, heat and power, and cattle feed) reduces the impact for climate change and fossil fuel depletion compared to the traditional biodiesel production; but the benefits for toxicity and eutrophication are limited.

Propionic acid produced by fermentation of glycerol leads to significant reduction of GHG emissions compared to fossil fuel alternatives; however the energy input is double and the contribution to eutrophication is significantly higher The LCA for the integration of butanol from prehydrolysate in a Canadian Kraft dissolving pulp mill shows that the carbon footprint of this butanol may be 5% lower compare to gasoline; but is not as low as corn butanol (23% lower than that of gasoline).

The majority of the LCA studies for the valorization of food waste have been focused on the environmental impacts on biogas or energy production, with only few on the synthesis of high value-added chemicals; hydroxymethylfurfural (HMF) has been listed as one of the top 10 bio-based chemicals by the US Department of Energy; the LCA of eight food waste valorization routes for the production of HMF shows that the most environmentally favorable option uses less polluting catalyst (AlCl3) and co-solvent (acetone), and provides the highest yield of HMF (27.9 Cmol%), metal depletion and toxicity impacts (marine ecotoxicity, freshwater toxicity, and human toxicity) were the categories with the highest values.

12.1.4 Biorefinery in the Pulp and Paper Industry

The pulp and paper industry is considered as the first industrialized biorefinery system; in this industrial process other co-products are produced including tall oil, rosin, vanillin, and lignosulfonates. Apart from these co-products; the system includes energy generation (in for of steam and electricity) to cover its internal energy demand; and it has the potential to feed heat and electricity to the grid. This industry has consolidated as the highest consumer of biomass; and uses not only wood as feedstock, it is capable of processing agricultural waste as bagasse, rice straw and corn stover. Other important features of this industry are a well-established logistic for biomass production, avoiding competition with food production for fertile land, and presenting higher biomass yields.

12.1.5 Examples

- The fully operational Blue Marble Energy company has multiple biorefineries located in Odessa, WA and Missoula, MT.
- Canada's first Integrated Biorefinery, developed on anaerobic digestion technology by Himark BioGas is located in Alberta. The biorefinery utilizes Source Separated Organics from the metro Edmonton region, open pen feedlot manure, and food processing waste.
- Chemrec's technology for black liquor gasification and production of second-generation biofuels such as biomethanol or BioDME is integrated with a host pulp mill and utilizes a major sulfate or sulfite process waste product as feedstock.
- Novamont has converted old petrochemical factories into biorefineries, producing protein, plastics, animal feed, lubricants, herbicides and elastomers from cardoon.
- C16 Biosciences produces synthetic palm oil from carbon-containing waste (i.e. food waste, glycerol) by means of yeast.
- MacroCascade aims to refine seaweed into food and fodder, and then products for healthcare, cosmetics, and fine chemicals industries. The side streams will be used for the production of fertilizer and biogas. Other seaweed biorefinery projects include MacroAlgaeBiorefinery (MAB4), SeaRefinery and SEAFARM.
- FUMI Ingredients produces foaming agents, heat-set gels and emulsifiers from micro-algae with the help of micro-organisms such as brewer's yeast and baker's yeast.
- The BIOCON platform is researching the processing of wood into various products. More precisely, their researchers are looking at transforming

lignin and cellulose into various products. Lignin for example can be transformed into phenolic components which can be used to make glue, plastics and agricultural products (e.g. crop protection). Cellulose can be transformed into clothes and packaging.

- In South Africa, Numbitrax LLC bought a Blume Biorefinery system for producing bioethanol as well as additional high-return offtake products from local and readily available resources such as the prickly pear cactus.
- Circular Organics (part of Kempen Insect Valley) grows black soldier fly larvae on waste from the agricultural and food industry (i.e. fruit and vegetable surplus, remaining waste from fruit juice and jam production). These larvae are used to produce protein, grease, and chitin. The grease is usable in the pharmaceutical industry (cosmetics, surfactants for shower gel), replacing other vegetable oils such as palm oil, or it can be used in fodder.
- Biteback Insect makes insect cooking oil, insect butter, fatty alcohols, insect frass protein and chitin from superworm (*Zophobas morio*).

12.2 PRIMARY METABOLISM IN BIOREFINERIES

Primary metabolism refers to the essential biochemical processes that support the growth and maintenance of an organism. In the context of biorefineries, understanding primary metabolism is crucial for optimizing the conversion of biomass into desired products. Primary metabolites, including carbohydrates, amino acids, lipids, and nucleotides, serve as building blocks for cellular structures and are pivotal for energy production.

12.2.1 Biomass Feedstock and Composition

In the context of energy production, biomass is matter from recently living (but now dead) organisms which is used for bioenergy production. Examples include wood, wood residues, energy crops, agricultural residues including straw, and organic waste from industry and households. Wood and wood residues is the largest biomass energy source today. Wood can be used as a fuel directly or processed into pellet fuel or other forms of fuels. Other plants can also be used as fuel, for instance maize, switchgrass, miscanthus and bamboo. The main waste feedstocks are wood waste, agricultural waste, municipal solid waste, and manufacturing waste. Upgrading raw biomass to higher grade fuels can be achieved by different methods, broadly classified as thermal, chemical, or biochemical.

The climate impact of bioenergy varies considerably depending on where biomass feedstocks come from and how they are grown. For example, burning wood for energy releases carbon dioxide. Those emissions can be significantly offset if the trees that were harvested are replaced by new trees in a well-managed forest, as the new trees will remove carbon dioxide from the air as they grow. However, the farming of biomass feedstocks can reduce biodiversity, degrade soils and take land out of food production. It may also consume water for irrigation and fertilisers.

Types and Uses

Different types of biomass are used for different purposes:

- Primary biomass sources that are appropriate for heat or electricity generation but not for transport include: wood, wood residues, wood pellets, agricultural residues, organic waste.
- Biomass that is processed into transport fuels can come from corn, sugar cane, and soy.

Biomass is categorized either as biomass harvested *directly for energy* (primary biomass), or as *residues and waste:* (secondary biomass).

Biomass Harvested Directly for Energy

The main biomass types harvested directly for energy is wood, some food crops and all perennial energy crops. One third of the global forest area of 4 billion hectares is used for wood production or other commercial purposes, and forests provide 85% of all biomass used for energy globally. In the EU, forests provide 60% of all biomass used for energy, with wood residues and waste being the largest source.

Woody biomass used for energy often consists of trees and bushes harvested for traditional cooking and heating purposes, particularly in developing countries, with 25 EJ per year used globally for these purposes. This practice is highly polluting. The World Health Organization (WHO) estimates that cooking-related pollution causes 3.8 million annual deaths. The United Nations Sustainable Development Goal 7 aims for the traditional use of biomass for cooking to be phased out by 2030. Short-rotation coppices and short-rotation forests are also harvested directly for energy, providing 4 EJ of energy, and are considered sustainable. The potential for these crops and perennial energy crops to provide at least 25 EJ annually by 2050 is estimated.

Food crops harvested for energy include sugar-producing crops (such as sugarcane), starch-producing crops (such as maize), and oil-producing crops (such as rapeseed). Sugarcane is a perennial crop, while corn and rapeseed are annual crops. Sugar- and starch-producing crops are used to make bioethanol, and oil-producing crops are used to make biodiesel. The United States is the largest producer of bioethanol, while the European Union is the largest producer of biodiesel. The global production of bioethanol and biodiesel provides 2.2 and 1.5 EJ of energy per year, respectively. Biofuel made from food crops harvested for energy is also known as "first-generation" or "traditional" biofuel and has relatively low emission savings.

The IPCC estimates that between 0.32 and 1.4 billion hectares of marginal land are suitable for bioenergy worldwide.

Biomass in the Form of Residues and Waste

Residues and waste are *by-products* from biological material harvested mainly for non-energy purposes. The most important by-products are wood residues, agricultural residues and municipal/industrial waste:

Wood residues are by-products from forestry operations or from the wood processing industry. Had the residues not been collected and used for bioenergy, they would have decayed (and therefore produced emissions) on the forest floor or in landfills, or been burnt (and produced emissions) at the side of the road in forests or outside wood processing facilities.

The by-products from forestry operations are called logging residues or forest residues, and consist of tree tops, branches, stumps, damaged or dying or dead trees, irregular or bent stem sections, thinnings (small trees that are cleared away in order to help the bigger trees grow large), and trees removed to reduce wildfire risk.

The extraction level of logging residues differ from region to region, but there is an increasing interest in using this feedstock, since the sustainable potential is large (15 EJ annually). 68% of the total forest biomass in the EU consists of wood stems, and 32% consists of stumps, branches and tops.

The by-products from the wood processing industry are called wood processing residues and consist of cut offs, shavings, sawdust, bark, and black liquor. Wood processing residues have a total energy content of 5.5 EJ annually. Wood pellets are mainly made from wood processing residues, and have a total energy content of 0.7 EJ. Wood chips are made from a combination of feedstocks, and have a total energy content of 0.8 EJ.

The energy content in agricultural residues used for energy is approximately 2 EJ. However, agricultural residues has a large untapped potential. The energy content in the global production of agricultural residues has been estimated to 78 EJ annually, with the largest share from straw (51 EJ). Others have estimated between 18 and 82 EJ. The use of agricultural residues and waste that is both sustainable and economically feasible is expected to increase to between 37 and 66 EJ in 2030.

Municipal waste produced 1.4 EJ and industrial waste 1.1 EJ. Wood waste from cities and industry also produced 1.1 EJ. The sustainable potential for wood waste has been estimated to 2–10 EJ. IEA recommends a dramatic increase in waste utilization to 45 EJ annually in 2050.

Biomass Conversion

Raw biomass can be upgraded into better and more practical fuel simply by compacting it (e.g. wood pellets), or by different conversions broadly classified as thermal, chemical, and biochemical. Biomass conversion reduces the transport costs as it is cheaper to transport high density commodities.

Thermal Conversion

Thermal upgrading produces solid, liquid or gaseous fuels, with heat as the dominant conversion driver. The basic alternatives are torrefaction, pyrolysis, and gasification, these are separated principally by how far the chemical reactions involved are allowed to proceed. The advancement of the chemical reactions is mainly controlled by how much oxygen is available, and the conversion temperature.

Torrefaction is a mild form of pyrolysis where organic materials are heated to 400–600 °F (200–300 °C) in a no-to-low oxygen environment. The heating process removes (via gasification) the parts of the biomass that has the lowest energy content, while the parts with the highest energy content remain. That is, approximately 30% of the biomass is converted to gas during the torrefaction process, while 70% remains, usually in the form of compacted pellets or briquettes. This solid product is water resistant, easy to grind, non-corrosive, and it contains approximately 85% of the original biomass energy. Basically the mass part has shrunk more than the energy part, and the consequence is that the calorific value of torrefied biomass increases significantly, to the extent that it can compete with coals used for electricity generation (steam/thermal coals). The energy density of the most common steam coals today is 22–26 GJ/t. There are other less common, more experimental or proprietary thermal processes that may offer benefits, such as hydrothermal upgrading (sometimes called "wet" torrefaction.) The hydrothermal upgrade path can be used for both low and high moisture content biomass, e.g. aqueous slurries.

Pyrolysis entails heating organic materials to 800–900 °F (400–500 °C) in the near complete absence of oxygen. Biomass pyrolysis produces fuels such as bio-oil, charcoal, methane, and hydrogen. Hydrotreating is used to process bio-oil (produced by fast pyrolysis) with hydrogen under elevated temperatures and pressures in the presence of a catalyst to produce renewable diesel, renewable gasoline, and renewable jet fuel.

Gasification entails heating organic materials to 1,400–1700 °F (800–900 °C) with injections of controlled amounts of oxygen and/or steam into the vessel to produce a carbon monoxide and hydrogen rich gas called synthesis gas or syngas. Syngas can be used as a fuel for diesel engines, for heating, and for generating electricity in gas turbines. It can also be treated to separate the hydrogen from the gas, and the hydrogen can be burned or used in fuel cells. The syngas can be further processed to produce liquid fuels using the Fischer-Tropsch synthesis process.

Chemical Conversion

A range of chemical processes may be used to convert biomass into other forms, such as to produce a fuel that is more practical to store, transport and use, or to exploit some property of the process itself. Many of these processes are based in large part on similar coal-based processes, such as the Fischer-Tropsch synthesis. A chemical conversion process known as transesterification is used for converting vegetable oils, animal fats, and greases into fatty acid methyl esters (FAME), which are used to produce biodiesel.

Biochemical Conversion

Biochemical processes have developed in nature to break down the molecules of which biomass is composed, and many of these can be harnessed. In most cases, microorganisms are used to perform the conversion. The processes are called anaerobic digestion, fermentation, and composting.

Fermentation converts biomass into bioethanol, and anaerobic digestion converts biomass into renewable natural gas (biogas). Bioethanol is used as a vehicle fuel. Renewable natural gas—also called biogas or biomethane—is produced in anaerobic digesters at sewage treatment plants and at dairy and livestock operations. It also forms in and may be captured from solid waste landfills. Properly treated renewable natural gas has the same uses as fossil fuel natural gas.

Climate Impacts

The climate impact of bioenergy varies considerably depending on where biomass feedstocks come from and how they are grown. For example, burning wood for energy releases carbon dioxide; those emissions can be significantly offset if the trees that were harvested are replaced by new trees in a well-managed forest, as the new trees will absorb carbon dioxide from the air as they grow. However, the establishment and cultivation of bioenergy crops can displace natural ecosystems, degrade soils, and consume water resources and synthetic fertilisers.

Approximately one-third of all wood used for traditional heating and cooking in tropical areas is harvested unsustainably. Bioenergy feedstocks typically require significant amounts of energy to harvest, dry, and transport; the energy usage for these processes may emit greenhouse gases. In some cases, the impacts of land-use change, cultivation, and processing can result in higher overall carbon emissions for bioenergy compared to using fossil fuels.

Use of farmland for growing biomass can result in less land being available for growing food. In the United States, around 10% of motor gasoline has been replaced by corn-based ethanol, which requires a significant proportion of the harvest. In Malaysia and Indonesia, clearing forests to produce palm oil for biodiesel has led to serious social and environmental effects, as these forests are critical carbon sinks and habitats for diverse species. Since photosynthesis captures only a small fraction of the energy in sunlight, producing a given amount of bioenergy requires a large amount of land compared to other renewable energy sources.

Short-Term vs. Long-Term Climate Benefits

Regarding the issue of climate consequences for modern bioenergy, IPCC states: "Life-cycle GHG emissions of modern bioenergy alternatives are usually lower than those for fossil fuels." Consequently, most of IPCC's GHG mitigation pathways include substantial deployment of bioenergy technologies.

Some research groups state that even if the European and North American forest carbon stock is increasing, it simply takes too long for harvested trees to grow back. Bioenergy from sources with high payback and parity times take a long time to have an impact on climate change mitigation. They therefore suggest that the EU should adjust its sustainability criteria so that only renewable energy with carbon payback times of less than 10 years is defined as sustainable, for instance wind, solar, biomass from wood residues and tree thinnings that would otherwise be burnt or decompose relatively fast, and biomass from short rotation coppicing (SRC).

The IPCC states: "While individual stands in a forest may be either sources or sinks, the forest carbon balance is determined by the sum of the net balance of all stands." IPCC also state that the only universally applicable approach to carbon accounting is the one that accounts for both carbon emissions and carbon removals (absorption) for managed lands (e.g. forest landscapes.) When the total is calculated, natural disturbances like fires and insect infestations are subtracted, and what remains is the human influence.

IEA Bioenergy state that an exclusive focus on the short-term make it harder to achieve efficient carbon mitigation in the long term, and compare investments in new bioenergy technologies with investments in other renewable energy technologies that only provide emission reductions after 2030, for instance the scaling-up of battery manufacturing or the development of rail infrastructure. Forest carbon emission avoidance strategies give a short-term mitigation benefit, but the long-term benefits from sustainable forestry activities provide ongoing forest product and energy resources.

Most of IPCC's GHG mitigation pathways include substantial deployment of bioenergy technologies. Limited or no bioenergy pathways leads to increased climate change or shifting bioenergy's mitigation load to other sectors. In addition, mitigation cost increases.

Carbon Accounting System Boundaries

Carbon positive scenarios are likely to be net emitters of CO_2, carbon negative projects are net absorbers of CO_2, while carbon neutral projects balance emissions and absorption equally. It is common to include *alternative* scenarios (also called "reference scenarios" or "counterfactuals") for comparison. The alternative scenarios range from scenarios with only modest changes compared to the existing project, all the way to radically different ones (i.e. forest protection or "no-bioenergy" counterfactuals.) Generally, the *difference* between scenarios is seen as the *actual* carbon mitigation potential of the scenarios.

In addition to the choice of alternative scenario, other choices has to be made as well. The so-called "system boundaries" determine *which* carbon emissions/ absorptions that will be included in the actual calculation, and which that will be excluded. System boundaries include temporal, spatial, efficiency-related and economic boundaries. For example, the actual carbon intensity of bioenergy varies with biomass production techniques and transportation lengths.

Temporal System Boundaries

The temporal boundaries define *when* to start and end carbon counting. Sometimes "early" events are included in the calculation, for instance carbon absorption going on in the forest before the initial harvest. Sometimes "late" events are included as well, for instance emissions caused by end-of-life activities for the infrastructure involved, e.g. demolition of factories. Since the emission and absorption of carbon related to a project or scenario changes with time, the net carbon emission can either be presented as *time-dependent* (for instance a curve which moves along a time axis), or as a *static value*; this shows *average* emissions calculated over a defined time period.

The *time-dependent* net emission curve will typically show high emissions at the beginning (if the counting starts when the biomass is harvested.) Alternatively, the starting point can be moved back to the planting event; in this case the curve can potentially move below zero (into carbon negative territory) if there is no carbon debt from land use change to pay back, and in addition more and more carbon is absorbed by the planted trees.

The emission curve then spikes upward at harvest. The harvested carbon is then being distributed into other carbon pools, and the curve moves in tandem with the *amount* of carbon that is moved into these new pools (Y axis), and the *time* it takes for the carbon to move out of the pools and return to the forest via the atmosphere (X axis). As described above, the carbon payback time is the time it takes for the harvested carbon to be returned to the forest, and the carbon parity time is the time it takes for the carbon stored in two competing scenarios to reach the same level.

The *static* carbon emission value is produced by calculating the *average* annual net emission for a specific time period. The specific time period can be the expected lifetime of the infrastructure involved (typical for life cycle assessments; LCA's), policy relevant time horizons inspired by the Paris agreement (for instance remaining time until 2030, 2050 or 2100), time spans based on different global warming potentials (GWP; typically 20 or 100 years), or other time spans. In the EU, a time span of 20 years is used when quantifying the net carbon effects of a land use change. Generally in legislation, the static number approach is preferred over the dynamic, time-dependent curve approach. The number is expressed as a so-called "emission factor" (net emission per produced energy unit, for instance kg CO_2e per GJ), or even simpler as an average greenhouse gas savings *percentage* for specific bioenergy pathways. The EU's published greenhouse gas savings percentages for specific bioenergy pathways used in the Renewable Energy Directive (RED) and other legal documents are based on life cycle assessments (LCA's).

Spatial System Boundaries

The spatial boundaries define "geographical" borders for carbon emission/absorption calculations. The two most common spatial boundaries for CO_2 absorption and emission in forests are 1.) along the edges of a particular forest stand and 2.) along the edges of a whole forest landscape, which include many forest stands of increasing age (the forest stands are harvested and replanted, one after the other, over as many years as there are stands.) A third option is the so-called increasing stand level carbon accounting method. The researcher has to decide whether to focus on the individual stand, an increasing number of stands, or the whole forest landscape. The IPCC recommends landscape-level carbon accounting.

Further, the researcher has to decide whether emissions from direct/indirect land use change should be included in the calculation. Most researchers include emissions from direct land use change, for instance the emissions caused by cutting down a forest in order to start some agricultural project there instead. The inclusion of indirect land use change effects is more controversial, as they are difficult to quantify accurately. Other choices involve defining the likely spatial boundaries of forests in the future.

Efficiency-Related System Boundaries

The efficiency-related boundaries define a range of fuel substitution efficiencies for different biomass-combustion pathways. Different supply chains emit different amounts of carbon per supplied energy unit, and different combustion facilities convert the chemical energy stored in different fuels to heat or electrical energy with different efficiencies. The researcher has to know about this and choose a realistic efficiency range for the different biomass-combustion paths under consideration.

The chosen efficiencies are used to calculate so-called "displacement factors" – single numbers that shows how efficient fossil carbon is substituted by biogenic carbon. If for instance 10 tonnes of carbon are combusted with an efficiency half that of a modern coal plant, only 5 tonnes of coal would actually be counted as displaced (displacement factor 0.5).

Generally, fuel burned in inefficient (old or small) combustion facilities gets assigned lower displacement factors than fuel burned in efficient (new or large) facilities, since more fuel has to be burned (and therefore more CO_2 released) in order to produce the same amount of energy.

The displacement factor varies with the carbon intensity of both the biomass fuel and the displaced fossil fuel. If or when bioenergy can achieve negative emissions (e.g. from afforestation, energy grass plantations and/or bioenergy with carbon capture and storage (BECCS), or if fossil fuel energy sources with higher emissions in the supply chain start to come online (e.g. because of fracking, or increased use of shale gas), the displacement factor will start to rise.

On the other hand, if or when new baseload energy sources with lower emissions than fossil fuels start to come online, the displacement factor will start to drop. Whether a displacement factor *change* is included in the calculation or not, depends on whether or not it is expected to take place within the time period covered by the relevant scenario's temporal system boundaries.

Economic System Boundaries

The economic boundaries define which market effects to include in the calculation, if any. Changed market conditions can lead to small or large changes in carbon emissions and absorptions from supply chains and forests, for instance changes in forest area as a response to changes in demand. Macroeconomic events/ policy changes can have impacts on forest carbon stock. Like with indirect land use changes, economic changes can be difficult to quantify however, so some researchers prefer to leave them out of the calculation.

System Boundary Impacts

The chosen system boundaries are very important for the calculated results. Shorter payback/parity times are calculated when fossil carbon intensity, forest growth rate and biomass conversion efficiency *increases*, or when the initial forest carbon stock and/or harvest level *decreases*. Shorter payback/parity times are also calculated when the researcher choose landscape level over stand level carbon accounting (if carbon accounting starts at the harvest rather than at the planting event.) Conversely, longer payback/parity times are calculated when carbon intensity, growth rate and conversion efficiency *decreases*, or when the initial carbon stock and/or harvest level *increases*, or the researcher choose stand level over landscape level carbon accounting. Critics argue that unrealistic system boundary choices are made, or that narrow system boundaries lead to misleading conclusions. Others argue that the wide range of results shows that there is too much leeway available and that the calculations therefore are useless for policy development. EU's Join Research Center agrees that different methodologies produce different results, but also argue that this is to be expected, since different researchers consciously or unconsciously choose different alternative scenarios/methodologies as a result of their ethical ideals regarding man's optimal relationship with nature. The ethical core of the sustainability debate should be made explicit by researchers, rather than hidden away.

Comparisons of GHG Emissions at the Point of Combustion

GHG emissions per produced energy unit at the point of combustion depend on moisture content in the fuel, chemical differences between fuels and conversion efficiencies. For example, raw biomass can have higher moisture content compared to some common coal types. When this is the case, more of the wood's inherent energy must be spent solely on evaporating moisture, compared to the drier coal, which means that the amount of CO_2 emitted per unit of produced heat will be higher.

Many biomass-only combustion facilities are relatively small and inefficient, compared to the typically much larger coal plants. Further, raw biomass (for instance wood chips) can have higher moisture content than coal (especially if the coal has been dried). When this is the case, more of the wood's inherent energy must be spent solely on evaporating moisture, compared to the drier coal, which means that the amount of CO_2 emitted per unit produced heat will be higher. This moisture problem can be mitigated by modern combustion facilities.

Forest biomass on average produces 10-16% more CO_2 than coal. However, focusing on gross emissions misses the point, what counts is the *net* climate effect from emissions *and* absorption, taken together. IEA Bioenergy concludes that the additional CO_2 from biomass "[...] is irrelevant if the biomass is derived from sustainably managed forests."

Climate Impacts Expressed as Varying with Time

The use of boreal stemwood harvested *exclusively* for bioenergy have a positive climate impact only in the long term, while the use of wood residues have a positive climate impact also in the short to medium term.

Short carbon payback/parity times are produced when the most realistic no-bioenergy scenario is a traditional forestry scenario where "good" wood stems are harvested for lumber production, and residues are burned or left behind in the forest or in landfills. The collection of such residues provides material which

"[...] would have released its carbon (via decay or burning) back to the atmosphere anyway (over time spans defined by the biome's decay rate) [...]." In other words, payback and parity times depend on the decay speed. The decay speed depends on a.) location (because decay speed is "[...] roughly proportional to temperature and rainfall [...]"), and b.) the thickness of the residues. Residues decay faster in warm and wet areas, and thin residues decay faster than thick residues. Thin residues in warm and wet temperate forests therefore have the fastest decay, while thick residues in cold and dry boreal forests have the slowest decay. If the residues instead are *burned* in the no-bioenergy scenario, e.g. outside the factories or at roadside in the forests, emissions are instant. In this case, parity times approach zero.

Like other scientists, the JRC staff note the high variability in carbon accounting results, and attribute this to different methodologies. In the studies examined, the JRC found carbon parity times of 0 to 400 years for stemwood harvested exclusively for bioenergy, depending on different characteristics and assumptions for both the forest/bioenergy system and the alternative fossil system, with the emission intensity of the displaced fossil fuels seen as the most important factor, followed by conversion efficiency and biomass growth rate/rotation time. Other factors relevant for the carbon parity time are the initial carbon stock and the existing harvest level; both higher initial carbon stock and higher harvest level means longer parity times. Liquid biofuels have high parity times because about half of the energy content of the biomass is lost in the processing.

Climate Impacts Expressed as Static Numbers

EU's Joint Research Centre has examined a number of bioenergy emission estimates found in literature, and calculated greenhouse gas savings percentages for bioenergy pathways in heat production, transportation fuel production and electricity production, based on those studies. The calculations are based on the attributional LCA accounting principle. It includes all supply chain emissions, from raw material extraction, through energy and material production and manufacturing, to end-of-life treatment and final disposal. It also includes emissions related to the production of the fossil fuels used in the supply chain. It excludes emission/absorption effects that takes place outside its system boundaries, for instance market related, biogeophysical (e.g. albedo), and time-dependent effects. The authors conclude that "[m]ost bio-based commodities release less GHG than fossil products along their supply chain; but the magnitude of GHG emissions vary greatly with logistics, type of feedstocks, land and ecosystem management, resource efficiency, and technology."

Because of the varied climate mitigation potential for different biofuel pathways, governments and organizations set up different certification schemes to ensure that biomass use is sustainable, for instance the RED (Renewable Energy Directive) in the EU and the ISO standard 13065 by the International Organization for Standardization. In the US, the RFS (Renewables Fuel Standard) limit the use of traditional biofuels and defines the minimum life-cycle GHG emissions that are acceptable. Biofuels are considered traditional if they achieve up to 20% GHG emission reduction compared to the petrochemical equivalent, advanced if they save at least 50%, and cellulosic if the save more than 60%.

The EU's Renewable Energy Directive (RED) states that the typical greenhouse gas emissions savings when replacing fossil fuels with wood pellets from forest residues for heat production varies between 69% and 77%, depending on transport distance: When the distance is between 0 and 2500 km, emission savings is 77%. Emission savings drop to 75% when the distance is between 2500 and 10 000 km, and to 69% when the distance is above 10 000 km. When stemwood is used, emission savings varies between 70% and 77%, depending on transport distance. When wood industry residues are used, savings varies between 79% and 87%.

Since the long payback and parity times calculated for some forestry projects is seen as a non-issue for energy crops (except in the cases mentioned above), researchers instead calculate *static* climate mitigation potentials for these crops, using LCA-based carbon accounting methods. A particular energy crop-based bioenergy project is considered carbon positive, carbon neutral or carbon negative based on the total amount of CO_2 equivalent emissions and absorptions accumulated throughout its entire lifetime: If emissions during agriculture, processing, transport and combustion are higher than what is absorbed (and stored) by the plants, both above and below ground, during the project's lifetime, the project is carbon positive. Likewise, if total absorption is higher than total emissions, the project is carbon negative. In other words, carbon negativity is possible when net carbon accumulation *more than compensates* for net lifecycle greenhouse gas emissions.

Typically, perennial crops sequester more carbon than annual crops because the root buildup is allowed to continue undisturbed over many years. Also, perennial crops avoid the yearly tillage procedures (plowing, digging) associated with growing annual crops. Tilling helps the soil microbe populations to decompose the available carbon, producing CO_2.

There is now (2018) consensus in the scientific community that "[...] the GHG [greenhouse gas] balance of perennial bioenergy crop cultivation will often be favourable [...]", also when considering the implicit direct and indirect land use changes.

Albedo and Evapotranspiration

Forests generally have a low albedo because the majority of the ultraviolet and visible spectrum is absorbed through photosynthesis. For this reason, the greater heat absorption by trees could offset some of the carbon benefits of afforestation (or offset the negative climate impacts of deforestation). In other words: The climate change mitigation effect of carbon sequestration by forests is partially counterbalanced in that reforestation can decrease the reflection of sunlight (albedo).

Environmental Impacts

The environmental impacts of biomass production need to be taken into account. For instance, in 2022, IEA stated that "bioenergy is an important pillar of decarbonisation in the energy transition as a near zero-emission fuel", and that "more efforts are needed to accelerate modern bioenergy deployment to get on track with the Net Zero Scenario [....] while simultaneously ensuring that bioenergy production does not incur negative social and environmental consequences."

Sustainable Forestry and Forest Protection

IPCC states that there is disagreement about whether the global forest is shrinking or not, and quote research indicating that tree cover has increased 7.1% between 1982 and 2016. The IPCC writes: "While above-ground biomass carbon stocks are estimated to be declining in the tropics, they are increasing globally due to increasing stocks in temperate and boreal forests [...]."

Old trees have a very high carbon absorption rate, and felling old trees means that this large potential for future carbon absorption is lost. There is also a loss of soil carbon due to the harvest operations.

Old trees absorb more CO_2 than young trees because of the larger leaf area in full-grown trees. However, the old forest (as a whole) will eventually stop absorbing CO_2 because CO_2 emissions from dead trees cancel out the remaining living trees' CO_2 absorption. The old forest (or forest stands) are also vulnerable for natural disturbances that produces CO_2. The IPCC found that "[...] landscapes with older forests have accumulated more carbon but their sink strength is diminishing, while landscapes with younger forests contain less carbon but they are removing CO_2 from the atmosphere at a much higher rate [...]."

The IPCC states that the net climate effect from conversion of unmanaged to managed forest can be positive or negative, depending on circumstances. The carbon stock is *reduced*, but since managed forests grow faster than unmanaged forests, more carbon is *absorbed*. Positive climate effects are produced if the harvested biomass is used efficiently. There is a tradeoff between the benefits of having a maximized forest carbon stock, not absorbing any more carbon, and the benefits of having a portion of that carbon stock "unlocked", and instead working as a renewable fossil fuel replacement tool, for instance in sectors which are difficult or expensive to decarbonize.

The "competition" between locked-away and unlocked forest carbon might be won by the unlocked carbon: "In the long term, using sustainably produced forest biomass as a substitute for carbon-intensive products and fossil fuels provides greater permanent reductions in atmospheric CO_2 than preservation does."

IEA Bioenergy writes: "forests managed for producing sawn timber, bioenergy and other wood products can make a greater contribution to climate change mitigation than forests managed for conservation alone." Three reasons are given:

- Reducing ability to act as a carbon sink when the forest matures.
- Wood products can replace other materials that emitted more GHGs during production.
- "Carbon in forests is vulnerable to loss through natural events such as insect infestations or wildfires".

Data from FAO show that most wood pellets are produced in regions dominated by sustainably managed forests, such as Europe and North America. Europe (including Russia) produced 54% of the world's wood pellets in 2019, and the forest carbon stock in this area increased from 158.7 to 172.4 Gt between 1990 and 2020.

In the EU, above-ground forest biomass increases with 1.3% per year on average, however the increase is slowing down because the forests are maturing.

United Kingdom Emissions Trading System allows operators of CO2 generating installations to apply zero emissions factor for the fraction used for non-energy purposes, while energy purposes (electricity generation, heating) require additional sustainability certification on the biomass used.

Biodiversity

Biomass production for bioenergy can have negative impacts on biodiversity. Oil palm and sugar cane are examples of crops that have been linked to reduced biodiversity. In addition, changes in biodiversity also impacts primary production which naturally effects decomposition and soil heterotrophic organisms.

Win-win scenarios (good for climate, good for biodiversity) include:

- Increased use of whole trees from coppice forests, increased use of thin forest residues from boreal forests with slow decay rates, and increased use of all kinds of residues from temperate forests with faster decay rates;
- Multi-functional bioenergy landscapes, instead of expansion of monoculture plantations; and
- Afforestation of former agricultural land with mixed or naturally regenerating forests.

Win-lose scenarios (good for the climate, bad for biodiversity) include afforestation on ancient, biodiversity-rich grassland ecosystems which were never forests, and afforestation of former agricultural land with monoculture plantations. Lose-win scenarios (bad for the climate, good for biodiversity) include natural forest expansion on former agricultural land. Lose-lose scenarios include increased use of thick forest residues like stumps from some boreal forests with slow decay rates, and conversion of natural forests into forest plantations.

Pollution

Other problems are pollution of soil and water from fertiliser/pesticide use, and emission of ambient air pollutants, mainly from open field burning of residues.

The traditional use of wood in cook stoves and open fires produces pollutants, which can lead to severe health and environmental consequences. However, a shift to modern bioenergy contribute to improved livelihoods and can reduce land degradation and impacts on ecosystem services. According to the IPCC, there is strong evidence that modern bioenergy have "large positive impacts" on air quality. Traditional bioenergy is inefficient and the phasing out of this energy source has both large health benefits and large economic benefits. When combusted in industrial facilities, most of the pollutants originating from woody biomass reduce by 97-99%, compared to open burning. Combustion of woody biomass produces lower amounts of particulate matter than coal for the same amount of electricity generated.

12.2.2 Fermentation Processes

Fermentation is a critical step in the conversion of biomass to biofuels and other products. Microbial fermentation utilizes the metabolic pathways of microorganisms, such as bacteria, yeast, and fungi, to convert fermentable sugars into desired end products. For example, yeast strains like *Saccharomyces cerevisiae* are widely used in bioethanol production due to their ability to efficiently ferment glucose into ethanol.

Optimizing fermentation processes is essential for maximizing yields and minimizing production costs. Factors such as temperature, pH, nutrient availability, and the presence of inhibitory compounds can significantly impact fermentation efficiency. Biorefineries must carefully monitor and control these parameters to ensure optimal microbial performance and product formation.

12.3 SECONDARY METABOLISM IN BIOREFINERIES

While primary metabolism is essential for growth and maintenance, secondary metabolism encompasses the biochemical pathways that produce specialized metabolites, which are not directly involved in growth but play critical roles in an organism's interaction with its environment. Secondary metabolites include a wide range of compounds, such as alkaloids, flavonoids, terpenoids, and phenolics, many of which possess valuable pharmaceutical and industrial applications.

12.3.1 Importance of Secondary Metabolites

The production of secondary metabolites in biorefineries offers significant opportunities for value addition. For instance, biorefineries can harness the biosynthetic pathways of microorganisms to produce high-value fine chemicals, pharmaceuticals, and nutraceuticals. The use of engineered microbial strains or plant cell cultures can enhance the yield and diversity of secondary metabolites, creating a lucrative market for bio-based products.

In the pharmaceutical industry, for example, secondary metabolites derived from plants and microorganisms have been utilized for their medicinal properties. Compounds like paclitaxel (from the Pacific yew tree) and artemisinin (from *Artemisia annua*) are crucial in cancer and malaria treatments, respectively. Biorefineries can play a pivotal role in producing these compounds sustainably through microbial fermentation or plant cell cultures, reducing reliance on traditional extraction methods that can lead to overharvesting and habitat destruction.

12.3.2 Challenges in Secondary Metabolite Production

Despite the potential of secondary metabolites, their production poses several challenges in biorefineries. One significant challenge is the regulatory complexity associated with the synthesis and commercialization of bio-based pharmaceuticals and chemicals. Regulatory agencies often impose stringent guidelines for the production, testing, and marketing of bio-based products, which can slow down innovation and increase production costs.

Additionally, the biosynthesis of secondary metabolites can be influenced by various factors, including the genetic makeup of the producing organism, the cultivation conditions, and the presence of precursor molecules. Achieving high yields of specific secondary metabolites often requires extensive metabolic engineering efforts, including the optimization of gene expression and the introduction of heterologous pathways. This complexity necessitates a multidisciplinary approach involving genetic engineering, biochemistry, and process optimization.

12.4 INDUSTRIAL PERSPECTIVES ON BIOREFINERIES

From an industrial perspective, biorefineries offer a pathway to sustainable production that aligns with growing consumer demand for eco-friendly products. As industries seek to reduce their carbon footprints and enhance sustainability, biorefineries provide a means to integrate renewable resources into the production of fuels, chemicals, and materials.

12.4.1 Economic Viability

The economic viability of biorefineries is a critical consideration for their successful implementation. To compete with fossil fuel-derived products, biorefineries must optimize their processes to reduce production costs while maintaining high product quality. This often involves investing in advanced technologies, such as biocatalysis, metabolic engineering, and process integration.

Moreover, the establishment of biorefineries requires significant capital investment and access to a reliable supply of biomass feedstock. Developing supply chains that ensure consistent feedstock availability is essential for maintaining operational efficiency and economic sustainability. Collaborative efforts among farmers, biomass suppliers, and biorefinery operators can foster the establishment of reliable supply chains and enhance the overall economic viability of biorefineries.

12.4.2 Policy and Regulatory Support

Government policies and regulatory frameworks play a crucial role in shaping the biorefinery landscape. Supportive policies, such as subsidies for renewable energy production, tax incentives for bio-based products, and research grants for biotechnological advancements, can stimulate investment in biorefineries and encourage innovation.

Regulatory agencies must also streamline the approval processes for bio-based products to facilitate market entry. This includes establishing clear guidelines for safety assessments, environmental impacts, and product labeling. By creating an enabling environment for biorefineries, governments can support the transition towards a bio-based economy.

12.5 FUTURE PROSPECTS AND CHALLENGES

The future of biorefineries is promising, with ongoing research and technological advancements poised to address existing challenges and unlock new opportunities. As the demand for sustainable products grows, biorefineries will play an increasingly vital role in resource utilization and environmental protection.

12.5.1 Advancements in Technology

Advancements in biotechnology, such as synthetic biology and metabolic engineering, will enhance the efficiency and productivity of biorefineries. These technologies can facilitate the design of microbial strains with optimized metabolic pathways, enabling higher yields of both primary and secondary metabolites.

Additionally, innovations in bioprocessing, including continuous fermentation and integrated downstream processing, can improve overall efficiency and reduce production costs.

12.5.2 Addressing Environmental Concerns

Biorefineries must also address environmental concerns associated with biomass production and conversion. Sustainable feedstock sourcing practices, such as utilizing agricultural residues and waste materials, can minimize land-use impacts and enhance the overall sustainability of biorefineries. Furthermore, incorporating waste-to-energy technologies and circular economy principles will help biorefineries minimize waste generation and optimize resource utilization.

12.6 CONCLUSION

Biorefineries represent a transformative approach to converting renewable resources into biofuels and valuable chemicals. By integrating primary and secondary metabolic processes, biorefineries can enhance resource utilization and promote sustainability in the production of energy and materials. While challenges related to feedstock variability, regulatory frameworks, and the production of secondary metabolites persist, advancements in technology and collaborative efforts among stakeholders can pave the way for a more sustainable and economically viable biorefinery landscape. As we move towards a bio-based economy, the successful implementation of biorefineries will be essential for meeting global energy demands while mitigating environmental impacts.

CHAPTER 13

SOLID-STATE FERMENTATION: CURRENT TRENDS AND FUTURE PROSPECTS

Solid-state fermentation (SSF) has emerged as an innovative bioprocess that utilizes solid substrates to cultivate microorganisms for the production of various valuable products, including enzymes, biofuels, and fermented foods. Unlike submerged fermentation, which relies on liquid media, SSF leverages the solid matrix of substrates, such as agricultural residues, to support microbial growth and metabolic activities. This method is particularly advantageous due to its lower water requirements, reduced energy consumption, and the potential for higher product yields. As industries increasingly seek sustainable and eco-friendly production methods, SSF has garnered significant attention.

13.1 UNDERSTANDING SOLID-STATE FERMENTATION

Solid state fermentation (SSF) is a biomolecule manufacturing process used in the food, pharmaceutical, cosmetic, fuel and textile industries. These biomolecules are mostly metabolites generated by microorganisms grown on a solid support selected for this purpose. This technology for the culture of microorganisms is an alternative to liquid or submerged fermentation, used predominantly for industrial purposes.

13.1.1 Processes

This process consists of depositing a solid culture substrate, such as rice or wheat bran, on flatbeds after seeding it with microorganisms; the substrate is then left in a temperature-controlled room for several days.

Liquid state fermentation is performed in tanks, which can reach 1,001 to 2,500 square metres (10,770 to 26,910 sq ft) at an industrial scale. Liquid culture is ideal for the growing of unicellular organisms such as bacteria or yeasts.

To achieve liquid aerobic fermentation, it is necessary to constantly supply the microorganism with oxygen, which is generally done via stirring the fermentation media. Accurately managing the synthesis of the desired metabolites requires regulating temperature, soluble oxygen, ionic strength, pH and control nutrients.

Applying this growing technique to filamentous fungi leads to difficulties. The fungus develops in its vegetative form, generating hyphae or multicellular ramous filaments, while a septum separates the cells. As this mycelium develops in a liquid environment, it generates abundant viscosity in the growing medium, reducing oxygen solubility, while stirring disrupts the cell network increasing cell mortality.

In nature, filamentous fungi grow on the ground, decomposing vegetal compounds under naturally ventilated conditions. Therefore, solid state fermentation enables the optimal development of filamentous fungi, allowing the mycelium to spread on the surface of solid compounds among which air can flow.

Solid state fermentation uses culture substrates with low water levels (reduced water activity), which is particularly appropriate for mould. The methods used to grow filamentous fungi using solid state fermentation allow the best reproduction of their natural environment. The medium is saturated with water but little of it is free-flowing. The solid medium comprises both the substrate and the solid support on which the fermentation takes place. The substrate used is generally composed of vegetal byproducts such as beet pulp or wheat bran.

At the beginning of the growth process, the substrates and solid culture compounds are non-soluble compounds composed of very large, biochemically complex molecules that the fungus will cut off to get essential C and N nutrients. To develop its natural substrate, the fungal organism sets forth its entire genetic potential to produce the metabolites necessary for its growth. The composition of the growth medium guides the microorganism's metabolism towards the production of enzymes that release bio-available single molecules such as sugars or amino acids by carving out macromolecules. Therefore, when selecting the components of the growth medium it is possible to guide the cells towards the production of the desired metabolite(s), mainly enzymes that transform polymers (cellulose, hemicellulose, pectins, proteins) into single moieties in a very efficient and cost-effective manner.

Compared to submerged fermentation processes, solid state fermentation is more cost-effective: smaller vessels, lower water consumption, reduced wastewater treatment costs and lower energy consumption (no need to heat up water, poor mechanical energy input due to smooth stirring).

Cultivating on heterogeneous substrates requires expertise to maintain optimal growth conditions. Air flow monitoring is key because it impacts temperature, oxygen supply and moisture. In order to maintain sufficient moisture content for the growth of filamentous fungus, waterlogged air is used and may require further addition of water. In most cases, solid state fermentation does not require a completely sterile environment as the initial sterilization of the fermentation substrate associated with the rapid colonization of the substrate by the fungous microorganism limits the development of the autochthonous flora.

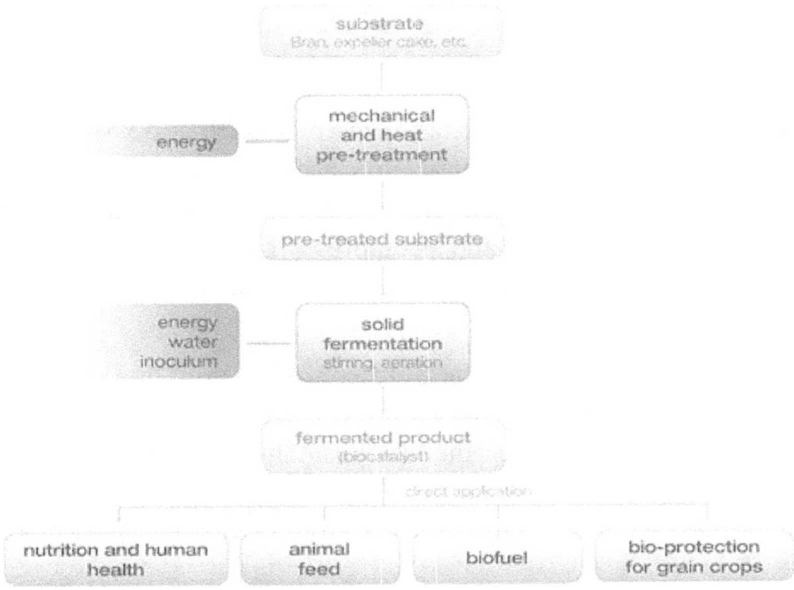

13.1.2 Uses

Traditional Food Production

Traditionally, SSF has been used in Asian countries to produce Koji using rice to manufacture alcoholic beverages such as Sake or Koji using soybean seeds. The latter produces sauces such as soy sauce or other foods. In Western countries, the traditional manufacturing process of many foods uses SSF. Examples include fermented bakery products such as bread or for the maturing of cheese. SSF is also widely used to prepare raw materials such as chocolate and coffee; typically cacao bean fermentation and coffee bean skin removal are SSF processes carried out under natural tropical conditions.

Enzyme Production

Enzymes and enzymatic complexes able to break down difficult-to-transform macromolecules such as cellulose, hemicelluloses, pectin and proteins. Solid state fermentation is well suited for the production of various enzymatic complexes composed of multiple enzymes. Enzymatic compounds generated by SSF find outlets in all sectors where digestibility, solubility or viscosity is needed.

This is why SSF enzymes are widely used in the following industries:

- Fruit and vegetable transformation (pectinases).
- Baking (hemicellulases).

- Animal feeding (hemicellulases and cellulases).
- Bio ethanol (cellulases and hemicellulases).
- Brewing and distilling (hemicellulases).

13.1.3 Advantages of SSF

The advantages of solid-state fermentation are manifold. One of the most significant benefits is its ability to utilize lignocellulosic biomass, which is abundant and often underutilized. By converting these materials into valuable products, SSF contributes to waste reduction and promotes a circular economy. Additionally, SSF generally requires less water than submerged fermentation, making it more suitable for regions facing water scarcity.

Furthermore, SSF can lead to higher product concentrations due to the concentrated nature of the solid substrate. This can result in improved yields of enzymes, bioactive compounds, and other metabolites. The solid-state environment can also enhance the production of secondary metabolites, such as antibiotics and flavor compounds, which are often sought after in the food and pharmaceutical industries.

13.2 CURRENT TRENDS IN SOLID-STATE FERMENTATION

13.2.1 Substrate Optimization

One of the current trends in solid-state fermentation is the optimization of substrate selection and preparation. Researchers are exploring various agro-industrial by-products as potential substrates, aiming to identify those that support optimal microbial growth and product formation. For instance, substrates rich in carbohydrates, such as sugarcane bagasse and fruit pomace, have gained attention for their potential to enhance enzyme production.

Additionally, substrate pretreatment methods, such as physical, chemical, and biological treatments, are being employed to improve the accessibility of nutrients to microorganisms. These pretreatment techniques can help break down complex lignocellulosic structures, facilitating microbial colonization and metabolic activity during fermentation.

13.2.2 Strain Selection and Genetic Engineering

The selection of appropriate microbial strains is crucial for the success of solid-state fermentation. Current research trends focus on identifying and isolating native strains of fungi, bacteria, and yeast that exhibit robust growth on solid substrates and high product yields. Filamentous fungi, such as *Aspergillus* and *Trichoderma*, are particularly popular due to their ability to degrade complex polysaccharides and produce valuable enzymes.

Moreover, advances in genetic engineering and synthetic biology have opened new avenues for strain improvement. By manipulating metabolic pathways, researchers can enhance the production of specific metabolites or improve the tolerance of microorganisms to adverse fermentation conditions, such as high temperatures or low pH. Engineered strains may also exhibit increased efficiency in substrate utilization, leading to higher overall yields.

13.2.3 Process Optimization and Scale-Up

Process optimization is another critical trend in solid-state fermentation. Researchers are investigating various fermentation parameters, including temperature, humidity, and inoculum size, to identify optimal conditions that maximize product yields. The use of response surface methodology and statistical modeling has facilitated the systematic evaluation of these parameters, allowing for more precise control of the fermentation process.

As the field of solid-state fermentation matures, scaling up from laboratory to industrial levels presents both opportunities and challenges. Current trends focus on developing scalable SSF systems that maintain the benefits of solid-state fermentation while accommodating larger volumes. Innovations in bioreactor design, such as the use of packed-bed reactors and rotary drum fermenters, are being explored to enhance mass transfer and aeration during fermentation.

13.3 FUTURE PROSPECTS OF SOLID-STATE FERMENTATION

13.3.1 Application in the Bioeconomy

The future prospects of solid-state fermentation are closely tied to the growing bioeconomy. As industries seek sustainable alternatives to fossil fuel-derived products, SSF can contribute to the production of biofuels, biochemicals, and bioplastics from renewable biomass. The ability to utilize agricultural waste as a feedstock aligns with the principles of circular economy, reducing environmental impact while creating value from waste materials.

Furthermore, SSF can play a significant role in the production of high-value products, such as enzymes and specialty chemicals. The demand for biocatalysts in various industries, including food, pharmaceuticals, and textiles, continues to rise, and solid-state fermentation offers a viable pathway for their sustainable production. By optimizing substrate selection and fermentation conditions, SSF can produce enzymes with specific properties tailored to meet industry needs.

13.3.2 Integration with Other Bioprocesses

Future advancements in solid-state fermentation may involve integrating SSF with other bioprocessing techniques to enhance product diversity and yield. For instance, coupling SSF with submerged fermentation can create hybrid systems that leverage the advantages of both methods. In such systems, solid substrates could be fermented in conjunction with liquid media, allowing for the simultaneous production of different metabolites.

Moreover, the integration of SSF with downstream processing techniques can streamline product recovery and purification. Innovations in extraction methods, such as supercritical fluid extraction and membrane filtration, can be employed to efficiently recover valuable metabolites from solid substrates, reducing waste and improving overall process efficiency.

13.3.3 Research and Technological Innovations

Ongoing research and technological innovations will play a pivotal role in shaping the future of solid-state fermentation. The application of omics technologies, such as genomics, proteomics, and metabolomics, can provide insights into microbial metabolism and product formation. These approaches can enhance our understanding of the complex interactions between microorganisms and solid substrates, leading to improved strain development and process optimization.

Additionally, advancements in automation and process monitoring technologies will enable real-time tracking of fermentation conditions, allowing for more precise control and optimization of SSF processes. The use of data analytics and machine learning can facilitate predictive modeling, helping researchers and industry stakeholders make informed decisions regarding substrate selection, strain optimization, and process scaling.

13.4 CHALLENGES AND CONSIDERATIONS

Despite its potential, solid-state fermentation faces several challenges that need to be addressed for successful implementation. One of the primary challenges is the variability of solid substrates, which can affect the reproducibility and consistency of fermentation processes. Variability in nutrient composition, moisture content, and particle size can influence microbial growth and product formation, necessitating careful control and standardization of substrate preparation.

Another challenge is the potential for contamination during fermentation, especially in large-scale operations. Contaminating microorganisms can compete with the desired strains for nutrients and may produce undesirable metabolites. Implementing robust sanitation protocols and employing biocontrol agents can help mitigate contamination risks.

13.5 CONCLUSION

Solid-state fermentation represents a promising and sustainable approach to microbial cultivation, with the potential to produce a wide array of valuable products. Current trends in substrate optimization, strain selection, and process optimization are paving the way for more efficient and effective SSF processes. As industries increasingly embrace the bioeconomy, solid-state fermentation is poised to play a vital role in producing biofuels, biochemicals, and other high-value products from renewable biomass.

Looking to the future, the integration of SSF with other bioprocesses, advancements in research and technology, and a focus on addressing existing challenges will drive the continued development and commercialization of solid-state fermentation.

By harnessing the power of microorganisms and solid substrates, SSF can contribute to a more sustainable and circular economy, creating value from waste and reducing the environmental impact of industrial production.

SECTION V:

FERMENTATION MICROBIOLOGY AND BIOTECHNOLOGY: TOOLS, MONITORING, AND CONTROL OF FERMENTATION PROCESSES

CHAPTER 14

FUNCTIONAL GENOMICS: CURRENT TRENDS, TOOLS, AND FUTURE PROSPECTS IN THE FERMENTATION AND PHARMACEUTICAL INDUSTRIES

Functional genomics is an interdisciplinary field that investigates the roles of genes and their interactions within biological systems. By employing various high-throughput technologies, functional genomics aims to elucidate the complex functions of genes and their products, particularly in the context of fermentation processes and pharmaceutical applications.

As the fermentation and pharmaceutical industries continue to evolve, the integration of functional genomics offers profound insights that can drive innovation, improve efficiency, and enhance product development.

14.1 UNDERSTANDING FUNCTIONAL GENOMICS

Functional genomics is a field of molecular biology that attempts to describe gene (and protein) functions and interactions. Functional genomics make use of the vast data generated by genomic and transcriptomic projects (such as genome sequencing projects and RNA sequencing).

Functional genomics focuses on the dynamic aspects such as gene transcription, translation, regulation of gene expression and protein–protein interactions, as opposed to the static aspects of the genomic information such as DNA sequence or structures. A key characteristic of functional genomics studies is their genome-wide approach to these questions, generally involving high-throughput methods rather than a more traditional "candidate-gene" approach.

14.1.1 Definition and Goals

In order to understand functional genomics it is important to first define function. In their paper Graur et al. define function in two possible ways. These are "selected effect" and "causal role". The "selected effect" function refers to the function for which a trait (DNA, RNA, protein etc.) is selected for. The "causal role" function refers to the function that a trait is sufficient and necessary for. Functional genomics usually tests the "causal role" definition of function.

The goal of functional genomics is to understand the function of genes or proteins, eventually all components of a genome. The term functional genomics is often used to refer to the many technical approaches to study an organism's genes and proteins, including the "biochemical, cellular, and/or physiological properties of each and every gene product" while some authors include the study of nongenic elements in their definition. Functional genomics may also include studies of natural genetic variation over time (such as an organism's development) or space (such as its body regions), as well as functional disruptions such as mutations.

The promise of functional genomics is to generate and synthesize genomic and proteomic knowledge into an understanding of the dynamic properties of an organism. This could potentially provide a more complete picture of how the genome specifies function compared to studies of single genes. Integration of functional genomics data is often a part of systems biology approaches.

14.1.2 Techniques and Applications

Functional genomics includes function-related aspects of the genome itself such as mutation and polymorphism (such as single nucleotide polymorphism (SNP) analysis), as well as the measurement of molecular activities. The latter comprise a number of "-omics" such as transcriptomics (gene expression), proteomics (protein production), and metabolomics. Functional genomics uses mostly multiplex techniques to measure the abundance of many or all gene products such as mRNAs or proteins within a biological sample. A more focused functional genomics approach might test the function of all variants of one gene and quantify the effects of mutants by using sequencing as a readout of activity. Together these measurement modalities endeavor to quantitate the various biological processes and improve our understanding of gene and protein functions and interactions.

At the DNA Level

Genetic Interaction Mapping

Systematic pairwise deletion of genes or inhibition of gene expression can be used to identify genes with related function, even if they do not interact physically. Epistasis refers to the fact that effects for two different gene knockouts may not be additive; that is, the phenotype that results when two genes are inhibited may be different from the sum of the effects of single knockouts.

DNA/Protein Interactions

Proteins formed by the translation of the mRNA (messenger RNA, a coded information from DNA for protein synthesis) play a major role in regulating gene expression. To understand how they regulate gene expression it is necessary to identify DNA sequences that they interact with. Techniques have been developed to identify sites of DNA-protein interactions. These include ChIP-sequencing, CUT&RUN sequencing and Calling Cards.

DNA Accessibility Assays

Assays have been developed to identify regions of the genome that are accessible. These regions of accessible chromatin are candidate regulatory regions. These assays include ATAC-seq, DNase-Seq and FAIRE-Seq.

At the RNA Level

Microarrays

Microarrays measure the amount of mRNA in a sample that corresponds to a given gene or probe DNA sequence. Probe sequences are immobilized on a solid surface and allowed to hybridize with fluorescently labeled "target" mRNA. The intensity of fluorescence of a spot is proportional to the amount of target sequence that has hybridized to that spot and therefore to the abundance of that mRNA sequence in the sample. Microarrays allow for the identification of candidate genes involved in a given process based on variation between transcript levels for different conditions and shared expression patterns with genes of known function.

SAGE

Serial Analysis of Gene Expression (SAGE) is a transcriptomic technique used by molecular biologists to produce a snapshot of the messenger RNA population in a sample of interest in the form of small tags that correspond to fragments of those transcripts. Several variants have been developed since, most notably a more robust version, LongSAGE, RL-SAGE and the most recent SuperSAGE. Many of these have improved the technique with the capture of longer tags, enabling more confident identification of a source gene.

Briefly, SAGE experiments proceed as follows:

- The mRNA of an input sample (e.g., a tumour) is isolated and a reverse transcriptase and biotinylated primers are used to synthesize cDNA from mRNA.
- The cDNA is bound to Streptavidin beads via interaction with the biotin attached to the primers, and is then cleaved using a restriction endonuclease called an anchoring enzyme (AE). The location of the cleavage site and thus the length of the remaining cDNA bound to the bead will vary for each individual cDNA (mRNA).
- The cleaved cDNA downstream from the cleavage site is then discarded, and the remaining immobile cDNA fragments upstream from

cleavage sites are divided in half and exposed to one of two adaptor oligonucleotides (A or B) containing several components in the following order upstream from the attachment site: 1) Sticky ends with the AE cut site to allow for attachment to cleaved cDNA; 2) A recognition site for a restriction endonuclease known as the tagging enzyme (TE), which cuts about 15 nucleotides downstream of its recognition site (within the original cDNA/mRNA sequence); 3) A short primer sequence unique to either adaptor A or B, which will later be used for further amplification via PCR.

- After adaptor ligation, cDNA are cleaved using TE to remove them from the beads, leaving only a short "tag" of about 11 nucleotides of original cDNA (15 nucleotides minus the 4 corresponding to the AE recognition site).
- The cleaved cDNA tags are then repaired with DNA polymerase to produce blunt end cDNA fragments.
- These cDNA tag fragments (with adaptor primers and AE and TE recognition sites attached) are ligated, sandwiching the two tag sequences together, and flanking adaptors A and B at either end. These new constructs, called ditags, are then PCR amplified using anchor A and B specific primers.
- The ditags are then cleaved using the original AE, and allowed to link together with other ditags, which will be ligated to create a cDNA concatemer with each ditag being separated by the AE recognition site.
- These concatemers are then transformed into bacteria for amplification through bacterial replication.
- The cDNA concatemers can then be isolated and sequenced using modern high-throughput DNA sequencers, and these sequences can be analysed with computer programs which quantify the recurrence of individual tags.

The output of SAGE is a list of short sequence tags and the number of times it is observed. Using sequence databases a researcher can usually determine, with some confidence, from which original mRNA (and therefore which gene) the tag was extracted. Statistical methods can be applied to tag and count lists from different samples in order to determine which genes are more highly expressed. For example, a normal tissue sample can be compared against a corresponding tumor to determine which genes tend to be more (or less) active.

The general goal of the technique is similar to the DNA microarray. However, SAGE sampling is based on sequencing mRNA output, not on hybridization of mRNA output to probes, so transcription levels are measured more quantitatively than by microarray. In addition, the mRNA sequences do not need to be known *a priori*, so genes or gene variants which are not known can be discovered. Microarray experiments are much cheaper to perform, so large-scale studies do not typically use SAGE. Quantifying gene expressions is more exact in SAGE because it involves directly counting the number of transcripts whereas spot intensities in microarrays fall in non-discrete gradients and are prone to background noise.

MicroRNAs, or miRNAs for short, are small (~22nt) segments of RNA which have been found to play a crucial role in gene regulation. One of the most commonly used methods for cloning and identifying miRNAs within a cell or tissue was developed in the Bartel Lab and published in a paper by Lau *et al.* (2001). Since then, several variant protocols have arisen, but most have the same basic format. The procedure is quite similar to SAGE: The small RNA are isolated, then linkers are added to each, and the RNA is converted to cDNA by RT-PCR. Following this, the linkers, containing internal restriction sites, are digested with the appropriate restriction enzyme and the sticky ends are ligated together into concatamers. Following concatenation, the fragments are ligated into plasmids and are used to transform bacteria to generate many copies of the plasmid containing the inserts. Those may then be sequenced to identify the miRNA present, as well as analysing expression levels of a given miRNA by counting the number of times it is present, similar to SAGE.

LongSAGE was a more robust version of the original SAGE developed in 2002 which had a higher throughput, using 20 µg of mRNA to generate a cDNA library of thousands of tags. Robust LongSage (RL-SAGE) Further improved on the LongSAGE protocol with the ability to generate a library with an insert size of 50 ng mRNA, much smaller than previous LongSAGE insert size of 2 µg mRNA and using a lower number of ditag polymerase chain reactions (PCR) to obtain a complete cDNA library.

SuperSAGE is a derivative of SAGE that uses the type III-endonuclease EcoP15I of phage P1, to cut 26 bp long sequence tags from each transcript's cDNA, expanding the tag-size by at least 6 bp as compared to the predecessor techniques SAGE and LongSAGE. The longer tag-size allows for a more precise allocation of the tag to the corresponding transcript, because each additional base increases the precision of the annotation considerably.

Like in the original SAGE protocol, so-called ditags are formed, using blunt-ended tags. However, SuperSAGE avoids the bias observed during the less random LongSAGE 20 bp ditag-ligation. By direct sequencing with high-throughput sequencing techniques (next-generation sequencing, i.e. pyrosequencing), hundred thousands or millions of tags can be analyzed simultaneously, producing very precise and quantitative gene expression profiles. Therefore, tag-based gene expression profiling also called "digital gene expression profiling" (DGE) can today provide most accurate transcription profiles that overcome the limitations of microarrays.

In the mid 2010s several techniques combined with Next Generation Sequencing were developed that employ the "tag" principle for "digital gene expression profiling" but without the use of the tagging enzyme. The "MACE" approach, (=Massive Analysis of cDNA Ends) generates tags somewhere in the last 1500 bps of a transcript. The technique does not depend on restriction enzymes anymore and thereby circumvents bias that is related to the absence or location of the restriction site within the cDNA. Instead, the cDNA is randomly fragmented and the 3'ends are sequenced from the 5' end of the cDNA molecule that carries the poly-A tail. The sequencing length of the tag can be freely chosen. Because of this, the tags can be assembled into contigs and the annotation of the tags can be drastically improved. Therefore, MACE is also use for the analyses of non-model organisms. In addition, the longer contigs can be screened for polymorphisms. As UTRs show a large number of polymorphisms between individuals, the MACE approach can be applied for allele determination, allele specific gene expression profiling and the search for molecular markers for breeding. In addition, the approach allows determining alternative polyadenylation of the transcripts. Because MACE does only require 3' ends of transcripts, even partly degraded RNA can be analyzed with less degradation dependent bias. The MACE approach uses unique molecular identifiers to allow for identification of PCR bias.

RNA Sequencing

RNA sequencing has taken over microarray and SAGE technology in recent years as noted in 2016, and has become the most efficient way to study transcription and gene expression.

This is typically done by next-generation sequencing. A subset of sequenced RNAs are small RNAs, a class of non-coding RNA molecules that are key regulators of transcriptional and post-transcriptional gene silencing, or RNA silencing. Next-generation sequencing is the gold standard tool for non-coding RNA discovery, profiling and expression analysis.

Massively Parallel Reporter Assays (MPRAs)

Massively parallel reporter assays is a technology to test the cis-regulatory activity of DNA sequences. MPRAs use a plasmid with a synthetic cis-regulatory element upstream of a promoter driving a synthetic gene such as Green Fluorescent Protein. A library of cis-regulatory elements is usually tested using MPRAs, a library can contain from hundreds to thousands of cis-regulatory elements. The cis-regulatory activity of the elements is assayed by using the downstream reporter activity. The activity of all the library members is assayed in parallel using barcodes for each cis-regulatory element. One limitation of MPRAs is that the activity is assayed on a plasmid and may not capture all aspects of gene regulation observed in the genome.

STARR-seq

STARR-seq (short for self-transcribing active regulatory region sequencing) is a method to assay enhancer activity for millions of candidates from arbitrary sources of DNA. It is used to identify the sequences that act as transcriptional enhancers in a direct, quantitative, and genome-wide manner.

In eukaryotes, transcription is regulated by sequence-specific DNA-binding proteins (transcription factors) associated with a gene's promoter and also by distant control sequences including enhancers. Enhancers are non-coding DNA sequences, containing several binding sites for a variety of transcription factors. They typically recruit transcriptional factors that modulate chromatin structure and directly interact with the transcription machinery placed at the promoter of a gene. Enhancers are able to regulate transcription of target genes in a cell type-specific manner, independent of their location or distance from the promoter of genes. In certain contexts, they can even regulate transcription of genes located in a different chromosome. However, the knowledge about enhancers so far has been limited to studies of a small number of enhancers, as they have been difficult to identify accurately at a genome-wide scale. Moreover, many regulatory elements function only in certain cell types and specific conditions.

Enhancer detection in Drosophila is an original methodology using random insertion of transposon-derived vector that encodes a reporter protein downstream of a minimal promoter. This approach allows to observe the expression of reporter in transgenic animals and provides information about nearby genes that are regulated by these sequences. The discovery and characterization of cell types along with genes involved in their determination have been significantly improved by the discovery of this technique.

During the past few years, post-genomic technologies, have displayed specific features of poised and active enhancers that have improved enhancer discovery. Development of new methods such as deep sequencing of DNase I hypersensitive sites (DNase-Seq), formaldehyde-assisted isolation of regulatory elements sequencing (FAIRE-Seq), chromatin immunoprecipitation followed by deep sequencing (ChIP-sequencing), and MNase-defined cistrome-Occupancy Analysis (MOA-seq), provide genome-wide enhancer predictions by enhancer-associated chromatin features.

However, DNase-seq and FAIRE-seq alone fail to provide a direct functional or quantitative readout of enhancer activity, so reporter assays that can deduce enhancer strength from the quantitative enrichment of reporter transcripts are needed to assess enhancer activity quantitatively. Yet, these assays are not high-throughput (High throughput biology), as it is impossible to conduct millions of tests required for identification of enhancers in a genome-wide manner. The development of STARR-seq attempts to circumvent this analytical barrier. Taking advantage of the knowledge that enhancers can work independently of their relative locations, candidate sequences are placed downstream of a minimal promoter, allowing the active enhancers to transcribe themselves. The strength of each enhancer is then reflected by its relative enrichment among cellular RNAs. Such a direct coupling of candidate sequences to enhancer activity enables the parallel evaluation of millions of DNA fragments from arbitrary sources.

Genomic DNA is randomly sheared and broken down to small fragments. Adaptors are ligated to size-selected DNA fragments. Next, adaptor-linked fragments are amplified and the PCR products are purified followed by placing candidate sequences downstream of a minimal promoter of screening vectors, giving them an opportunity to transcribe themselves. Candidate cells are then transfected with reporter library and cultured. Thereafter, total RNAs are extracted and poly-A RNAs isolated. Using reverse transcription method, cDNAs are produced, amplified and then candidate fragments are used for high-throughput paired end sequencing. Sequence reads are mapped to the reference genome and computational processing of data is carried out.

Applying this technology to Drosophila genome, Arnold et al. found 96% of the non-repetitive genome with at least 10-fold coverage. Authors discovered that most identified enhancers (55.6%) were placed within introns, particularly in the first intron and intergenic regions. 4.5% of enhancers were located at transcription start sites (TSS), suggesting that these enhancers can start transcription and also improve transcription from a distant TSS. The strongest enhancers were near housekeeping genes such as enzymes or component of the cytoskeleton and developmental regulators such as the transcription factors. The strongest enhancer was located within the intron of the transcription factor zfh1. This transcription factor regulates neuropeptide expression and growth of larval neuromuscular junctions in Drosophila. The ribosomal protein genes were the only class of genes with poor enhancers ranking. Moreover, authors demonstrated that many genes are regulated by several independent active enhancers even in a single cell type. Furthermore, gene expression levels on average were correlated with the sum of the enhancer strengths per gene, supporting direct link between gene expression and enhancer activity.

Applying this technology to the characterization and discovery of regulatory variant alleles, Vockley et al. characterized the effects of human genetic variation on non-coding regulatory element function, measuring the activity of 100 putative enhancers captured directly from the genomes of 95 members of a study cohort. This approach enables the functional fine-mapping of causal regulatory variants in regions of high linkage disequilibrium identified by eQTL analyses. This approach provides a general path forward to identify perturbations in gene regulatory elements that contribute to complex phenotypes.

STARR-seq has been used to measure the regulatory activity of DNA fragments that have been enriched for sites occupied by specific transcription factors. Cloning ChIP DNA libraries generated from chromatin immunoprecipitation of the glucocorticoid receptor into STARR-seq enabled genome-scale quantification of glucocorticoid-induced enhancer activity. This approach is useful for measuring the differences in enhancer activity between sites that are bound by the same transcription factor.

By combining traditional approach with high-throughput sequencing technology and highly specialized bio-computing methods, STARR-seq is able to detect enhancers in a quantitative and genome-wide manner. The study of gene regulation and their responsible pathways in the genome during normal development and also in disease can be very demanding. Therefore, applying STARR-seq to many cell types across organisms supports identifying cell type-specific gene regulatory elements and practically assesses non-coding mutations causing disease. Recently, a related approach coupling capture of regions of interest to STARR-seq technique have been developed and extensively validated in mammalian cell lines.

- **Advantages:**
  - A quantitative genome-wide assay for enhancer detection.
  - Applicable technique for screening arbitrary sources of DNA in any cell type or tissue that allows adequate introduction of reporter constructs.
  - A method with high detection rate (>99%) by employing pair-end sequencing, even for sequences that contain transcript-destabilizing elements.
  - Technique to evaluate the strength of enhancers quantitatively, and identify endogenously silenced enhancers by integrating them into a chromosomal context.

Perturb-seq

Perturb-seq (also known as CRISP-seq and CROP-seq) refers to a high-throughput method of performing single cell RNA sequencing (scRNA-seq) on pooled genetic perturbation screens. Perturb-seq combines multiplexed CRISPR mediated gene inactivations with single cell RNA sequencing to assess comprehensive gene expression phenotypes for each perturbation. Inferring a gene's function by applying genetic perturbations to knock down or knock out a gene and studying the resulting phenotype is known as reverse genetics. Perturb-seq is a reverse genetics approach that allows for the investigation of phenotypes at the level of the transcriptome, to elucidate gene functions in many cells, in a massively parallel fashion.

The Perturb-seq protocol uses CRISPR technology to inactivate specific genes and DNA barcoding of each guide RNA to allow for all perturbations to be pooled together and later deconvoluted, with assignment of each phenotype to a specific guide RNA. Droplet-based microfluidics platforms (or other cell sorting and separating techniques) are used to isolate individual cells, and then scRNA-seq is performed to generate gene expression profiles for each cell. Upon completion of the protocol, bioinformatics analyses are conducted to associate each specific cell and perturbation with a transcriptomic profile that characterizes the consequences of inactivating each gene.

In the December 2016 issue of the Cell journal, two companion papers were published that each introduced and described this technique. A third paper describing a conceptually similar approach (termed CRISP-seq) was also published in the same issue. In October 2016, the CROP-seq method for single-cell CRISPR screening was presented in a preprint on bioRxiv and later published in the Nature Methods journal. While each paper shared the core principles of combining CRISPR mediated perturbation with scRNA-seq, their experimental, technological and analytical approaches differed in several aspects, to explore distinct biological questions, demonstrating the broad utility of this methodology. For example, the CRISPR-seq paper demonstrated the feasibility of *in vivo* studies using this technology, and the CROP-seq protocol facilitates large screens by providing a vector that makes the guide RNA itself readable (rather than relying on expressed barcodes), which allows for single-step guide RNA cloning. A June 2022 paper in *Cell* published results from one of the first genome-scale Perturb-seq screens, which uncovered new perturbations that promote chromosomal instability as well as variations in the expression of mitochondrially encoded transcripts in response to different forms of mitochondrial stress.

Pooled CRISPR libraries that enable gene inactivation can come in the form of either knockout or interference. Knockout libraries perturb genes through double stranded breaks that prompt the error prone non-homologous end joining repair pathway to introduce disruptive insertions or deletions. CRISPR interference (CRISPRi) on the other hand utilizes a catalytically inactive nuclease to physically block RNA polymerase, effectively preventing or halting transcription. Perturb-seq has been utilized with both the knockout and CRISPRi approaches in the Dixit et al. paper and the Adamson et al. paper, respectively.

Pooling all guide RNAs into a single screen relies on DNA barcodes that act as identifiers for each unique guide RNA. There are several commercially available pooled CRISPR libraries including the guide barcode library used in the study by Adamson et al. CRISPR libraries can also be custom made using tools for sgRNA design, many of which are listed on the CRISPR/cas9 tools Wikipedia page.

The sgRNA expression vector design will depend largely on the experiment performed but requires the following central components:

- Promoter.
- Restriction sites.
- Primer Binding Sites.
- sgRNA.
- Guide Barcode.
- **Reporter Gene:**
  - Fluorescent gene: vectors are often constructed to include a gene encoding a fluorescent protein, such that successfully transduced cells can be visually and quantitatively assessed by their expression.
  - Antibiotic resistance gene: similar to fluorescent markers, antibiotic resistance genes are often incorporated into vectors to allow for selection of successfully transduced cells.
- **CRISPR-Associated Endonuclease:** Cas9 or other CRISPR-associated endonucleases such as Cpf1 must be introduced to cells that do not endogenously express them. Due to the large size of these genes, a two-vector system can be used to express the endonuclease separately from the sgRNA expression vector.

Cells are typically transduced with a Multiplicity of Infection (MOI) of 0.4 to 0.6 lentiviral particles per cell to maximize the likelihood of obtaining the most cells which contain a single guide RNA. If the effects of simultaneous perturbations are of interest, a higher MOI may be applied to increase the amount of transduced cells with more than one guide RNA. Selection for successfully transduced cells is then performed using a fluorescence assay or an antibiotic assay, depending on the reporter gene used in the expression vector.

After successfully transduced cells have been selected for, isolation of single cells is needed to conduct scRNA-seq. Perturb-seq and CROP-seq have been performed using droplet-based technology for single cell isolation, while the closely related CRISP-seq was performed with a microwell-based approach. Once cells have been isolated at the single cell level, reverse transcription, amplification and sequencing takes place to produce gene expression profiles for each cell. Many scRNA-seq approaches incorporate unique molecular identifiers (UMIs) and cell barcodes during the reverse transcription step to index individual RNA molecules and cells, respectively. These additional barcodes serve to help quantify RNA transcripts and to associate each of the sequences with their cell of origin.

Read alignment and processing are performed to map quality reads to a reference genome. Deconvolution of cell barcodes, guide barcodes and UMIs enables the association of guide RNAs with the cells that contain them, thus allowing the gene expression profile of each cell to be affiliated with a particular perturbation. Further downstream analyses on the transcriptional profiles will depend entirely on the biological question of interest. T-distributed Stochastic Neighbor Embedding (t-SNE) is a commonly used machine learning algorithm to visualize the high-dimensional data that results from scRNA-seq in a 2-dimensional scatterplot. The authors who first performed Perturb-seq developed an in-house computational framework called MIMOSCA that predicts the effects of each perturbation using a linear model and is available on an open software repository.

Perturb-seq makes use of current technologies in molecular biology to integrate a multi-step workflow that couples high-throughput screening with complex phenotypic outputs. When compared to alternative methods used for gene knockdowns or knockouts, such as RNAi, zinc finger nucleases or transcription activator-like effector nucleases (TALENs), the application of CRISPR-based perturbations enables more specificity, efficiency and ease of use. Another advantage of this protocol is that while most screening approaches can only assay for simple phenotypes, such as cellular viability, scRNA-seq allows for a much richer phenotypic readout, with quantitative measurements of gene expression in many cells simultaneously. Perturb-seq can therefore combine the high throughput of forward genetics, in terms of the number of genetic perturbations, with the rich phenotype dimension of reverse genetics.

However, while a large and comprehensive amount of data can be a benefit, it can also present a major challenge. Single cell RNA expression readouts are known to produce 'noisy' data, with a significant number of false positives. Both the large size and noise that is associated with scRNA-seq will likely require new and powerful computational methods and bioinformatics pipelines to better make sense of the resulting data. Another challenge associated with this protocol is the creation of large scale CRISPR libraries. The preparation of these extensive libraries depends upon a comparative increase in the resources required to culture the massive numbers of cells that are needed to achieve a successful screen of many perturbations.

In parallel to these single-cell methods, other approaches have been developed to reconstruct genetic pathways using whole-organism RNA-sequencing. These methods use a single aggregate statistic, called the transcriptome-wide epistasis coefficient, to guide pathway reconstruction. In contrast with the statistical framework of the methods described above, this coefficient may be more robust to noise and is intuitively interpretable in terms of Batesonian epistasis. This approach was used to identify a new state in the life cycle of the nematode *C. elegans*.

Perturb-seq or other conceptually similar protocols can be used to address a broad scope of biological questions and the applications of this technology will likely grow over time. Three papers on this topic, published in the December 2016 issue of the Journal Cell, demonstrated the utility of this method by applying it to the investigation of several distinct biological functions. In the paper, "Perturb-Seq: Dissecting Molecular Circuits with Scalable Single-Cell RNA Profiling of Pooled Genetic Screens", the authors used Perturb-seq to conduct knockouts of transcription factors related to the immune response in hundreds of thousands of cells to investigate the cellular consequences of their inactivation. They also explored the effects of transcription factors on cell states in the context of the cell cycle. In the study led by UCSF, "A Multiplexed Single-Cell CRISPR Screening Platform Enables Systematic Dissection of the Unfolded Protein Response" the researchers suppressed multiple genes in each cell to study the unfolded protein response (UPR) pathway. With a similar methodology, but using the term CRISP-seq instead of Perturb-seq, the paper "Dissecting Immune Circuits by Linking CRISPR-Pooled Screens with Single-Cell RNA-Seq" performed a proof of concept experiment by using the technique to probe regulatory pathways related to innate immunity in mice. Lethality of each perturbation and epistasis analyses in cells with multiple perturbations was also investigated in these papers. Perturb-seq has so far been used with very few perturbations per experiment, but it can theoretically be scaled up to address the whole genome. Finally, the October 2016 preprint and subsequent paper demonstrate the bioinformatic reconstruction of the T cell receptor signaling pathway in Jurkat cells based on CROP-seq data.

Recently, the Perturb-seq (CROP-seq) workflow has been adapted to enable genome-scale CRISPRi (CRISPR interference) screens in Jurkat cells at single-cell resolution. The first-of-its-kind genome-scale CRISPRi screen was conducted to verify factors involved in TCR signaling pathways. In more detail, a guide RNA library targeting 18,595 human genes was utilized for CRISPR-based gene knockdowns in Jurkat cells expressing the dCas9-KRAB fusion endonuclease. In total, one million Jurkat cells were processed for single-cell RNA sequencing allowing transcriptomic readouts of a final list of 374 marker genes involved in TCR signaling. The bioinformatic analysis confirmed more than 70 known activators and repressors of TCR signaling cascades, hence showcasing the potential of Perturb-seq (CROP-seq) screens to support translational research.

While these publications used these protocols for answering complex biological questions, this technology can also be used as a validation assay to ensure the specificity of any CRISPR based knockdown or knockout; the expression levels of the target genes as well as others can be measured with single cell resolution in parallel, to detect whether the perturbation was successful and to assess the experiment for off target effects. Furthermore, these protocols make it possible to perform perturbation screens in heterogeneous tissues, while obtaining cell type specific gene expression responses.

At the Protein Level

Yeast Two-Hybrid System

A yeast two-hybrid screening (Y2H) tests a "bait" protein against many potential interacting proteins ("prey") to identify physical protein–protein interactions. This system is based on a transcription factor, originally GAL4, whose separate DNA-binding and transcription activation domains are both required in order for the protein to cause transcription of a reporter gene. In a Y2H screen, the "bait" protein is fused to the binding domain of GAL4, and a library of potential "prey" (interacting) proteins is recombinantly expressed in a vector with the activation domain. In vivo interaction of bait and prey proteins in a yeast cell brings the activation and binding domains of GAL4 close enough together to result in expression of a reporter gene. It is also possible to systematically test a library of bait proteins against a library of prey proteins to identify all possible interactions in a cell.

MS and AP/MS

Mass spectrometry (MS) can identify proteins and their relative levels, hence it can be used to study protein expression. When used in combination with affinity purification, mass spectrometry (AP/MS) can be used to study protein complexes, that is, which proteins interact with one another in complexes and in which ratios. In order to purify protein complexes, usually a "bait" protein is tagged with a specific protein or peptide that can be used to pull out the complex from a complex mix. The purification is usually done using an antibody or a compound that binds to the fusion part. The proteins are then digested into short peptide fragments and mass spectrometry is used to identify the proteins based on the mass-to-charge ratios of those fragments.

Deep Mutational Scanning

In deep mutational scanning, every possible amino acid change in a given protein is first synthesized. The activity of each of these protein variants is assayed in parallel using barcodes for each variant. By comparing the activity to the wild-type protein, the effect of each mutation is identified. While it is possible to assay every possible single amino-acid change due to combinatorics two or more concurrent mutations are hard to test. Deep mutational scanning experiments have also been used to infer protein structure and protein-protein interactions. Deep Mutational Scanning is an example of a multiplexed assays of variant effect (MAVEs), a family of methods that involve mutagenesis of a DNA-encoded protein or regulatory element followed by a multiplexed assay for some aspect of function. MAVEs enable the generation of 'variant effect maps' characterizing aspects of the function of every possible single nucleotide change in a gene or functional element of interest.

Mutagenesis and Phenotyping

An important functional feature of genes is the phenotype caused by mutations. Mutants can be produced by random mutations or by directed mutagenesis, including site-directed mutagenesis, deleting complete genes, or other techniques.

Knock-Outs (Gene Deletions)

Gene function can be investigated by systematically "knocking out" genes one by one. This is done by either deletion or disruption of function (such as by insertional mutagenesis) and the resulting organisms are screened for phenotypes that provide clues to the function of the disrupted gene. Knock-outs have been produced for whole genomes, i.e. by deleting all genes in a genome. For essential genes, this is not possible, so other techniques are used, e.g. deleting a gene while expressing the gene from a plasmid, using an inducible promoter, so that the level of gene product can be changed at will (and thus a "functional" deletion achieved).

Site-Directed Mutagenesis

Site-directed mutagenesis is used to mutate specific bases (and thus amino acids). This is critical to investigate the function of specific amino acids in a protein, e.g. in the active site of an enzyme.

RNAi

RNA interference (RNAi) is a biological process in which RNA molecules are involved in sequence-specific suppression of gene expression by double-stranded RNA, through translational or transcriptional repression. Historically, RNAi was known by other names, including *co-suppression*, *post-transcriptional gene silencing* (PTGS), and *quelling*. The detailed study of each of these seemingly different processes elucidated that the identity of these phenomena were all actually RNAi. Andrew Fire and Craig C. Mello shared the 2006 Nobel Prize in Physiology or Medicine for their work on RNAi in the nematode worm *Caenorhabditis elegans*, which they published in 1998. Since the discovery of RNAi and its regulatory potentials, it has become evident that RNAi has immense potential in suppression of desired genes. RNAi is now known as precise, efficient, stable and better than antisense therapy for gene suppression. Antisense RNA produced intracellularly by an expression vector may be developed and find utility as novel therapeutic agents.

Two types of small ribonucleic acid (RNA) molecules, microRNA (miRNA) and small interfering RNA (siRNA), are central to components to the RNAi pathway. Once mRNA is degraded, post-transcriptional silencing occurs as protein translation is prevented. Transcription can be inhibited via the pre-transcriptional silencing mechanism of RNAi, through which an enzyme complex catalyzes DNA methylation at genomic positions complementary to complexed siRNA or miRNA. RNAi has an important role in defending cells against parasitic nucleotide sequences (e.g., viruses or transposons) and also influences development of organisms.

The RNAi pathway is a naturally occurring process found in many eukaryotes and animal cells. It is initiated by the enzyme Dicer, which cleaves long double-stranded RNA (dsRNA) molecules into short double-stranded fragments of approximately 21 to 23 nucleotide siRNAs. Each siRNA is unwound into two single-stranded RNAs (ssRNAs), the passenger (sense) strand and the guide (antisense) strand. The passenger strand is then cleaved by the protein Argonaute 2 (Ago2). The passenger strand is degraded and the guide strand is incorporated into the RNA-induced silencing complex (RISC). The RISC assembly then binds and degrades the target mRNA. Specifically, this is accomplished when the guide strand pairs with a complementary sequence in a mRNA molecule and induces cleavage by Ago2, a catalytic component of the RISC. In some organisms, this process spreads systemically, despite the initially limited molar concentrations of siRNA.

RNAi is a valuable research tool, both in cell culture and in living organisms, because synthetic dsRNA introduced into cells can selectively and robustly induce suppression of specific genes of interest. RNAi may be used for large-scale screens that systematically shut down each gene (and the subsequent proteins it codes for) in the cell, which can help to identify the components necessary for a particular cellular process or an event such as cell division. The pathway is also used as a practical tool for food, medicine and insecticides.

RNAi is an RNA-dependent gene silencing process that is controlled by RISC and is initiated by short double-stranded RNA molecules in a cell's cytoplasm, where they interact with the catalytic RISC component Argonaute. When the dsRNA is exogenous (coming from infection by a virus with an RNA genome or laboratory manipulations), the RNA is imported directly into the cytoplasm and cleaved to short fragments by Dicer. The initiating dsRNA can also be endogenous (originating in the cell), as in pre-microRNAs expressed from RNA-coding genes in the genome. The primary transcripts from such genes are first processed to form the characteristic stem-loop structure of pre-miRNA in the nucleus, then exported to the cytoplasm. Thus, the two dsRNA pathways, exogenous and endogenous, converge at the RISC.

Exogenous dsRNA initiates RNAi by activating the ribonuclease protein Dicer, which binds and cleaves dsRNAs in plants, or short hairpin RNAs (shRNAs) in humans, to produce double-stranded fragments of 20–25 base pairs with a 2-nucleotide overhang at the 3′ end. Bioinformatics studies on the genomes of multiple organisms suggest this length maximizes target-gene specificity and minimizes non-specific effects. These short double-stranded fragments are called siRNAs. These siRNAs are then separated into single strands and integrated into an active RISC, by RISC-Loading Complex (RLC). RLC includes Dicer-2 and R2D2, and is crucial to unite Ago2 and RISC. TATA-binding protein-associated factor 11 (TAF11) assembles the RLC by facilitating Dcr-2-R2D2 tetramerization, which increases the binding affinity to siRNA by 10-fold. Association with TAF11 would convert the R2-D2-Initiator (RDI) complex into the RLC. R2D2 carries tandem double-stranded RNA-binding domains to recognize the thermodynamically stable terminus of siRNA duplexes, whereas Dicer-2 the other less stable extremity. Loading is asymmetric: the MID domain of Ago2 recognizes the thermodynamically stable end of the siRNA. Therefore, the "passenger" (sense) strand whose 5′ end is discarded by MID is ejected, while the saved "guide" (antisense) strand cooperates with AGO to form the RISC.

After integration into the RISC, siRNAs base-pair to their target mRNA and cleave it, thereby preventing it from being used as a translation template. Differently from siRNA, a miRNA-loaded RISC complex scans cytoplasmic mRNAs for potential complementarity. Instead of destructive cleavage (by Ago2), miRNAs rather target the 3′ untranslated region (UTR) regions of mRNAs where they typically bind with imperfect complementarity, thus blocking the access of ribosomes for translation.

Exogenous dsRNA is detected and bound by an effector protein, known as RDE-4 in *C. elegans* and R2D2 in *Drosophila*, that stimulates Dicer activity. The mechanism producing this length specificity is unknown and this protein only binds long dsRNAs.

In *C. elegans* this initiation response is amplified through the synthesis of a population of 'secondary' siRNAs during which the Dicer-produced initiating or 'primary' siRNAs are used as templates. These 'secondary' siRNAs are structurally distinct from Dicer-produced siRNAs and appear to be produced by an RNA-dependent RNA polymerase (RdRP).

MicroRNAs (miRNAs) are genomically encoded non-coding RNAs that help regulate gene expression, particularly during development. The phenomenon of RNAi, broadly defined, includes the endogenously induced gene silencing effects of miRNAs as well as silencing triggered by foreign dsRNA. Mature miRNAs are structurally similar to siRNAs produced from exogenous dsRNA, but before reaching maturity, miRNAs must first undergo extensive post-transcriptional modification. A miRNA is expressed from a much longer RNA-coding gene as a primary transcript known as a *pri-miRNA* which is processed, in the cell nucleus, to a 70-nucleotide stem-loop structure called a *pre-miRNA* by the microprocessor complex. This complex consists of an RNase

III enzyme called Drosha and a dsRNA-binding protein DGCR8. The dsRNA portion of this pre-miRNA is bound and cleaved by Dicer to produce the mature miRNA molecule that can be integrated into the RISC complex; thus, miRNA and siRNA share the same downstream cellular machinery. First, viral encoded miRNA was described in Epstein–Barr virus (EBV). Thereafter, an increasing number of microRNAs have been described in viruses. VIRmiRNA is a comprehensive catalogue covering viral microRNA, their targets and anti-viral miRNAs.

siRNAs derived from long dsRNA precursors differ from miRNAs in that miRNAs, especially those in animals, typically have incomplete base pairing to a target and inhibit the translation of many different mRNAs with similar sequences. In contrast, siRNAs typically base-pair perfectly and induce mRNA cleavage only in a single, specific target. In *Drosophila* and *C. elegans*, miRNA and siRNA are processed by distinct Argonaute proteins and Dicer enzymes.

Three prime untranslated regions (3′UTRs) of mRNAs often contain regulatory sequences that post-transcriptionally cause RNAi. Such 3′-UTRs often contain both binding sites for miRNAs as well as for regulatory proteins. By binding to specific sites within the 3′-UTR, miRNAs can decrease gene expression of various mRNAs by either inhibiting translation or directly causing degradation of the transcript. The 3′-UTR also may have silencer regions that bind repressor proteins that inhibit the expression of a mRNA.

The 3′-UTR often contains microRNA response elements (MREs). MREs are sequences to which miRNAs bind. These are prevalent motifs within 3′-UTRs. Among all regulatory motifs within the 3′-UTRs (e.g. including silencer regions), MREs make up about half of the motifs.

As of 2023, the miRBase web site, an archive of miRNA sequences and annotations, listed 28,645 entries in 271 biologic species. Of these, 1,917 miRNAs were in annotated human miRNA loci. miRNAs were predicted to have an average of about four hundred target mRNAs (affecting expression of several hundred genes). Friedman et al. estimate that >45,000 miRNA target sites within human mRNA 3′UTRs are conserved above background levels, and >60% of human protein-coding genes have been under selective pressure to maintain pairing to miRNAs.

Direct experiments show that a single miRNA can reduce the stability of hundreds of unique mRNAs. Other experiments show that a single miRNA may repress the production of hundreds of proteins, but that this repression often is relatively mild (less than 2-fold).

The effects of miRNA dysregulation of gene expression seem to be important in cancer. For instance, in gastrointestinal cancers, nine miRNAs have been identified as epigenetically altered and effective in down regulating DNA repair enzymes.

The effects of miRNA dysregulation of gene expression also seem to be important in neuropsychiatric disorders, such as schizophrenia, bipolar disorder, major depression, Parkinson's disease, Alzheimer's disease and autism spectrum disorders.

Exogenous dsRNA is detected and bound by an effector protein, known as RDE-4 in *C. elegans* and R2D2 in *Drosophila*, that stimulates Dicer activity. This protein only binds long dsRNAs, but the mechanism producing this length specificity is unknown. This RNA-binding protein then facilitates the transfer of cleaved siRNAs to the RISC complex.

The active components of an RNA-induced silencing complex (RISC) are endonucleases called Argonaute proteins, which cleave the target mRNA strand complementary to their bound siRNA. As the fragments produced by Dicer are double-stranded, they could each in theory produce a functional siRNA. However, only one of the two strands, which is known as the *guide strand*, binds Argonaute and directs gene silencing. The other *anti-guide strand* or *passenger strand* is degraded during RISC activation. Although it was first believed that an ATP-dependent helicase separated these two strands, the process proved to be ATP-independent and performed directly by the protein components of RISC. However, an *in vitro* kinetic analysis of RNAi in the presence and absence of ATP showed that ATP may be required to unwind and remove the cleaved mRNA strand from the RISC complex after catalysis. The guide strand tends to be the one whose 5′ end is less stably paired to its complement, but strand selection is unaffected by the direction in which Dicer cleaves the dsRNA before RISC incorporation. Instead, the R2D2 protein may serve as the differentiating factor by binding the more-stable 5′ end of the passenger strand.

The structural basis for binding of RNA to the Argonaute protein was examined by X-ray crystallography of the binding domain of an RNA-bound Argonaute. Here, the phosphorylated 5′ end of the RNA strand enters a conserved basic surface pocket and makes contacts through a divalent cation (an atom with two positive charges) such as magnesium and by aromatic stacking (a process that allows more than one atom to share an electron by passing it back and forth) between the 5′ nucleotide in the siRNA and a conserved tyrosine residue. This site is thought to form a nucleation site for the binding of the siRNA to its mRNA target. Analysis of the inhibitory effect of mismatches in either the 5' or 3' end of the guide strand has demonstrated that the 5' end of the guide strand is likely responsible for matching and binding the target mRNA, while the 3' end is responsible for physically arranging target mRNA into a cleavage-favorable RISC region.

It is not understood how the activated RISC complex locates complementary mRNAs within the cell. Although the cleavage process has been proposed to be linked to translation, translation of the mRNA target is not essential for RNAi-mediated degradation. Indeed, RNAi may be more effective against mRNA targets that are not translated. Argonaute proteins are localized to specific regions in the cytoplasm called P-bodies (also cytoplasmic bodies or GW bodies), which are regions with high rates of mRNA decay; miRNA activity is also clustered in P-bodies. Disruption of P-bodies decreases the efficiency of RNAi, suggesting that they are a critical site in the RNAi process.

Components of the RNAi pathway are used in many eukaryotes in the maintenance of the organization and structure of their genomes. Modification of histones and associated induction of heterochromatin formation serves to downregulate genes pre-transcriptionally; this process is referred to as RNA-induced transcriptional silencing (RITS), and is carried out by a complex of proteins called the RITS complex. In fission yeast this complex contains Argonaute, a chromodomain protein Chp1, and a protein called Tas3 of unknown function. As a consequence, the induction and spread of heterochromatic regions requires the Argonaute and RdRP proteins. Indeed, deletion of these genes in the fission yeast *S. pombe* disrupts histone methylation and centromere formation, causing slow or stalled anaphase during cell division. In some cases, similar processes associated with histone modification have been observed to transcriptionally upregulate genes.

The mechanism by which the RITS complex induces heterochromatin formation and organization is not well understood. Most studies have focused on the mating-type region in fission yeast, which may not be representative of activities in other genomic regions/organisms. In maintenance of existing heterochromatin regions, RITS forms a complex with siRNAs complementary to the local genes and stably binds local methylated histones, acting co-transcriptionally to degrade any nascent pre-mRNA transcripts that are initiated by RNA polymerase. The formation of such a heterochromatin region, though not its maintenance, is Dicer-dependent, presumably because Dicer is required to generate the initial complement of siRNAs that target subsequent transcripts. Heterochromatin maintenance has been suggested to function as a self-reinforcing feedback loop, as new siRNAs are formed from the occasional nascent transcripts by RdRP for incorporation into local RITS complexes. The relevance of observations from fission yeast mating-type regions and centromeres to mammals is not clear, as heterochromatin maintenance in mammalian cells may be independent of the components of the RNAi pathway.

The type of RNA editing that is most prevalent in higher eukaryotes converts adenosine nucleotides into inosine in dsRNAs via the enzyme adenosine deaminase (ADAR). It was originally proposed in 2000 that the RNAi and A→I RNA editing pathways might compete for a common dsRNA substrate. Some pre-miRNAs do undergo A→I RNA editing and this mechanism may regulate the processing and expression of mature miRNAs. Furthermore, at least one mammalian ADAR can sequester siRNAs from RNAi pathway components. Further support for this model comes from studies on ADAR-null *C. elegans* strains indicating that A→I RNA editing may counteract RNAi silencing of endogenous genes and transgenes.

Organisms vary in their ability to take up foreign dsRNA and use it in the RNAi pathway. The effects of RNAi can be both systemic and heritable in plants and *C. elegans*, although not in *Drosophila* or mammals. In plants, RNAi is thought to propagate by the transfer of siRNAs between cells through plasmodesmata (channels in the cell walls that enable communication and transport). Heritability comes from methylation of promoters targeted by RNAi; the new methylation pattern is copied in each new generation of the cell. A broad general distinction between plants and animals lies in the targeting of endogenously produced miRNAs; in plants, miRNAs are usually perfectly or nearly perfectly complementary to their target genes and induce direct mRNA cleavage by RISC, while animals' miRNAs tend to be more divergent in sequence and induce translational repression. This translational effect may be produced by inhibiting the interactions of translation initiation factors with the mRNA's polyadenine tail.

Some eukaryotic protozoa such as *Leishmania major* and *Trypanosoma cruzi* lack the RNAi pathway entirely. Most or all of the components are also missing in some fungi, most notably the model organism *Saccharomyces cerevisiae*. The presence of RNAi in other budding yeast species such as *Saccharomyces castellii* and *Candida albicans*, further demonstrates that inducing two RNAi-related proteins from *S. castellii* facilitates RNAi in *S. cerevisiae*. That certain ascomycetes and basidiomycetes are missing RNAi pathways indicates that proteins required for RNA silencing have been lost independently from many fungal lineages, possibly due to the evolution of a novel pathway with similar function, or to the lack of selective advantage in certain niches.

Gene expression in prokaryotes is influenced by an RNA-based system similar in some respects to RNAi. Here, RNA-encoding genes control mRNA abundance or translation by producing a complementary RNA that anneals to an mRNA. However these regulatory RNAs are not generally considered to be analogous to miRNAs because the Dicer enzyme is not involved. It has been suggested that CRISPR interference systems in prokaryotes are analogous to eukaryotic RNAi systems, although none of the protein components are orthologous.

RNAi is a vital part of the immune response to viruses and other foreign genetic material, especially in plants where it may also prevent the self-propagation of transposons. Plants such as *Arabidopsis thaliana* express multiple Dicer homologs that are specialized to react differently when the plant is exposed to different viruses. Even before the RNAi pathway was fully understood, it was known that induced gene silencing in plants could spread throughout the plant in a systemic effect and could be transferred from stock to scion plants via grafting. This phenomenon has since been recognized as a feature of the plant immune system which allows the entire plant to respond to a virus after an initial localized encounter. In response, many plant viruses have evolved elaborate mechanisms to suppress the RNAi response. These include viral proteins that bind short double-stranded RNA fragments with single-stranded overhang ends, such as those produced by Dicer. Some plant genomes also express endogenous siRNAs in response to infection by specific types of bacteria. These effects may be part of a generalized response to pathogens that downregulates any metabolic process in the host that aids the infection process.

Although animals generally express fewer variants of the Dicer enzyme than plants, RNAi in some animals produces an antiviral response. In both juvenile and adult *Drosophila*, RNAi is important in antiviral innate immunity and is active against pathogens such as Drosophila X virus. A similar role in immunity may operate in *C. elegans*, as Argonaute proteins are upregulated in response to viruses and worms that overexpress components of the RNAi pathway are resistant to viral infection.

The role of RNAi in mammalian innate immunity is poorly understood, and relatively little data is available. However, the existence of viruses that encode genes able to suppress the RNAi response in mammalian cells may be evidence in favour of an RNAi-dependent mammalian immune response, although this hypothesis has been challenged as poorly substantiated. Evidence for the existence of a functional antiviral RNAi pathway in mammalian cells has been presented.

Other functions for RNAi in mammalian viruses also exist, such as miRNAs expressed by the herpes virus that may act as heterochromatin organization triggers to mediate viral latency.

Endogenously expressed miRNAs, including both intronic and intergenic miRNAs, are most important in translational repression and in the regulation of development, especially on the timing of morphogenesis and the maintenance of undifferentiated or incompletely differentiated cell types such as stem cells. The role of endogenously expressed miRNA in downregulating gene expression was first described in *C. elegans* in 1993. In plants this function was discovered when the "JAW microRNA" of *Arabidopsis* was shown to be involved in the regulation of several genes that control plant shape.

In plants, the majority of genes regulated by miRNAs are transcription factors; thus miRNA activity is particularly wide-ranging and regulates entire gene networks during development by modulating the expression of key regulatory genes, including transcription factors as well as F-box proteins. In many organisms, including humans, miRNAs are linked to the formation of tumors and dysregulation of the cell cycle. Here, miRNAs can function as both oncogenes and tumor suppressors.

Based on parsimony-based phylogenetic analysis, the most recent common ancestor of all eukaryotes most likely already possessed an early RNAi pathway; the absence of the pathway in certain eukaryotes is thought to be a derived characteristic. This ancestral RNAi system probably contained at least one Dicer-like protein, one Argonaute, one PIWI protein, and an RNA-dependent RNA polymerase that may also have played other cellular roles. A large-scale comparative genomics study likewise indicates that the eukaryotic crown group already possessed these components, which may then have had closer functional associations with generalized RNA degradation systems such as the exosome. This study also suggests that the RNA-binding Argonaute protein family, which is shared among eukaryotes, most archaea, and at least some bacteria (such as *Aquifex aeolicus*), is homologous to and originally evolved from components of the translation initiation system.

Gene knockdown is a method used to reduce the expression of an organism's specific genes. This is accomplished by using the naturally occurring process of RNAi. This gene knockdown technique uses a double-stranded siRNA molecule that is synthesized with a sequence complementary to the gene of interest. The RNAi cascade begins once the Dicer enzyme starts to process siRNA. The end result of the process leads to degradation of mRNA and destroys any instructions needed to build certain proteins. Using this method, researchers are able to decrease (but not completely eliminate) the expression of a targeted gene. Studying the effects of this decrease in expression may show the physiological role or impact of the targeted gene products.

Extensive efforts in computational biology have been directed toward the design of successful dsRNA reagents that maximize gene knockdown but minimize "off-target" effects. Off-target effects arise when an introduced RNA has a base sequence that can pair with and thus reduce the expression of multiple genes. Such problems occur more frequently when the dsRNA contains repetitive sequences. It has been estimated from studying the genomes of humans, *C. elegans* and *S. pombe* that about 10% of possible siRNAs have substantial off-target effects. A multitude of software tools have been developed implementing algorithms for the design of general mammal-specific, and virus-specific siRNAs that are automatically checked for possible cross-reactivity.

Depending on the organism and experimental system, the exogenous RNA may be a long strand designed to be cleaved by Dicer, or short RNAs designed to serve as siRNA substrates. In most mammalian cells, shorter RNAs are used because long double-stranded RNA molecules induce the mammalian interferon response, a form of innate immunity that reacts nonspecifically to foreign genetic material.

Mouse oocytes and cells from early mouse embryos lack this reaction to exogenous dsRNA and are therefore a common model system for studying mammalian gene-knockdown effects. Specialized laboratory techniques have also been developed to improve the utility of RNAi in mammalian systems by avoiding the direct introduction of siRNA, for example, by stable transfection with a plasmid encoding the appropriate sequence from which siRNAs can be transcribed, or by more elaborate lentiviral vector systems allowing the inducible activation or deactivation of transcription, known as *conditional RNAi*.

The technique of knocking down genes using RNAi therapeutics has demonstrated success in randomized controlled clinical studies. These medications are a growing class of siRNA-based drugs that decrease the expression of proteins encoded by certain genes. To date, five RNAi medications have been approved by regulatory authorities in the US and Europe: patisiran (2018), givosiran (2019), lumasiran (2020), inclisiran (2020 in Europe with anticipated US approval in 2021), and vutrisiran (2022).

While all of the current regulatory body approved RNAi therapeutics focus on diseases that originate in the liver, additional medications under investigation target a host of disease areas including cardiovascular diseases, bleeding disorders, alcohol use disorders, cystic fibrosis, gout, carcinoma, and eye disorders.

Patisiran is the first double stranded siRNA-based medication approved in 2018 and developed by Alnylam Pharmaceuticals. Patisiran uses the RNAi cascade to suppress the gene that codes for TTR (transthryetin). Mutations in this gene may cause the misfolding of a protein responsible for hereditary ATTR amyloidosis. To achieve therapeutic response, patisiran is encased by a lipid nanoparticle membrane that facilitates crossover into the cytoplasm. Once inside the cell, the siRNA begins processing by the enzyme Dicer. Patisiran is administered by a healthcare professional through an intravenous infusion with dosing based on body weight. Warnings and precautions include risk of infusion-related reactions and reduced vitamin A levels (serum).

In 2019, the FDA and EMA approved givosiran for the treatment of adults with acute hepatic porphyria (AHP). The FDA also granted givosiran a breakthrough therapy designation, priority review designation, and orphan drug designation for the treatment of acute hepatic porphyria (AHP) in November 2019. By 2020, givosiran received EMA approval. Givosiran is an siRNA that breaks down aminolevulinic acid synthase 1 (ALAS1) mRNA in the liver. Breaking down ALAS1 mRNA prevents toxins (responsible for neurovisceral attacks and AHP disease) such as aminolevulinic acid (ALA) and porphobilinogen (PBG) from accumulating. To facilitate entry into the cytoplasm, givosiran uses GalNAc ligands and enters into liver cells. The medication is administered subcutaneously by a healthcare professional with dosing based on body weight. Warnings and precautions include risk of anaphylactic reactions, hepatic toxicity, renal toxicity and injection site reactions.

Lumasiran was approved as a siRNA-based medication in 2020 for use in both the European Union and the United States. This medication is used for the treatment of primary hyperoxaluria type 1 (PH1) in pediatric and adult populations. The drug is designed to reduce hepatic oxalate production and urinary oxalate levels through RNAi by targeting hydroxyacid oxidase 1 (HAO1) mRNA for breakdown. Lowering HAO1 enzyme levels reduces the oxidation of glycolate to glyoxylate (which is a substrate for oxalate). Lumasiran is administered subcutaneously by a healthcare professional with dosing based on body weight. Data from randomized controlled clinical trials indicate that the most common adverse reaction that was reported was injection site reactions. These reactions were mild and were present in 38 percent of patients treated with lumasiran.

In 2022, the FDA and EMA approved vutrisiran for the treatment of adults with hereditary transthyretin mediated amyloidosis with polyneuropathy stage 1 or 2. Vutrisiran is designed to break down the mRNA that codes for transthyretin.

Other investigational drugs using RNAi that are being developed by pharmaceutical companies such as Arrowhead Pharmaceuticals, Dicerna, Alnylam Pharmaceuticals, Amgen, and Sylentis. These medications cover a variety of targets via RNAi and diseases.

Currently, both miRNA and SiRNA are currently chemically synthesized and so, are legally categorized inside EU and in USA as "simple" medicinal products. But as bioengineered siRNA (BERAs) are in development, these would be classified as biological medicinal products, at least in EU. The development of the BERAs technology raises the question of the categorization of drugs having the same mechanism of action but being produced chemically or biologically. This lack of consistency should be addressed.

To achieve the clinical potential of RNAi, siRNA must be efficiently transported to the cells of target tissues. However, there are various barriers that must be fixed before it can be used clinically. For example, "naked" siRNA is susceptible to several obstacles that reduce its therapeutic efficacy. Additionally, once siRNA has entered the bloodstream, naked RNA can be degraded by serum nucleases and can stimulate the innate immune system. Due to its size and highly polyanionic (containing negative charges at several sites) nature, unmodified siRNA molecules cannot readily enter the cells through the cell membrane. Therefore, artificial or nanoparticle encapsulated siRNA must be used. If siRNA is transferred across the cell membrane, unintended toxicities can occur if therapeutic doses are not optimized, and siRNAs can exhibit off-target effects (e.g. unintended downregulation of genes with partial sequence complementarity). Even after entering the cells, repeated dosing is required since their effects are diluted at each cell division. In response to these potential issues and barriers, two approaches help facilitate siRNA delivery to target cells: lipid nanoparticles and conjugates.

Lipid nanoparticles (LNPs) are based on liposome-like structures that are typically made of an aqueous center surrounded by a lipid shell. A subset of liposomal structures used for delivery drugs to tissues rest in large unilamellar vesicles (LUVs) which may be 100 nm in size. LNP delivery mechanisms have become an increasing source of encasing nucleic acids and may include plasmids, CRISPR and mRNA.

The first approved use of lipid nanoparticles as a drug delivery mechanism began in 2018 with the siRNA drug patisiran, developed by Alnylam Pharmaceuticals. Dicerna Pharmaceuticals, Persomics, Sanofi and Sirna Therapeutics also worked to bring RNAi therapies to market.

Other recent applications include two FDA approved COVID-19 vaccines: mRNA-1273, developed by Moderna. and BNT162b, developed by a collaboration between Pfizer and BioNtech. These two vaccines use lipid nanoparticles to deliver antigen mRNA. Encapsulating the mRNA molecule in lipid nanoparticles was a critical breakthrough for producing viable mRNA vaccines, solving a number of key technical barriers in delivering the mRNA molecule into the host cell as distributed through apolipoprotein E (apoE) in the low-density lipoprotein receptor (LDLR). In December 2020, Novartis announced that positive results from phase III efficacy studies deemed inclisiran was a treatment for heterozygous familial hypercholesterolemia (HeFH) and atherosclerotic cardiovascular disease (ASCVD).

In addition to LNPs, RNAi therapeutics have targeted delivery through siRNA conjugates (e.g., GalNAc, carbohydrates, peptides, aptamers, antibodies). Therapeutics using siRNA conjugates have been developed for rare or genetic diseases such as acute hepatic porphyria (AHP), hemophilia, primary hyperoxaluria (PH) and hereditary ATTR amyloidosis as well as other cardiometabolic diseases such as hypertension and non-alcoholic steatohepatitis (NASH).

RNAi has been used for a variety of other applications including food, crops and insecticides. The use of the RNAi pathway has developed numerous products such as foods like Arctic apples, nicotine-free tobacco, decaffeinated coffee, nutrient fortified vegetation and hypoallergenic crops. The emerging use of RNAi has the potential to develop many other products for future use.

Antiviral treatment is one of the earliest proposed RNAi-based medical applications, and two different types have been developed. The first type is to target viral RNAs. Many studies have shown that targeting viral RNAs can suppress the replication of numerous viruses, including HIV, HPV, hepatitis A, hepatitis B, influenza virus, respiratory syncytial virus (RSV), SARS coronavirus (SARS-CoV), adenovirus and measles virus. The other strategy is to block the initial viral entries by targeting the host cell genes. For example, suppression of chemokine receptors (CXCR4 and CCR5) on host cells can prevent HIV viral entry.

While traditional chemotherapy can effectively kill cancer cells, lack of specificity for discriminating normal cells and cancer cells in these treatments usually cause severe side effects. Numerous studies have demonstrated that RNAi can provide a more specific approach to inhibit tumor growth by targeting cancer-related genes (i.e., oncogene).

It has also been proposed that RNAi can enhance the sensitivity of cancer cells to chemotherapeutic agents, providing a combinatorial therapeutic approach with chemotherapy. Another potential RNAi-based treatment is to inhibit cell invasion and migration.

Compared with chemotherapy or other anti-cancer drugs, there are a lot of advantages of siRNA drug. SiRNA acts on the post-transcriptional stage of gene expression, so it does not modify or change DNA in a deleterious effect. SiRNA can also be used to produce a specific response in a certain type of way, such as by downgrading suppression of gene expression. In a single cancer cell, siRNA can cause dramatic suppression of gene expression with just several copies. This happens by silencing cancer-promoting genes with RNAi, as well as targeting an mRNA sequence.

RNAi drugs treat cancer by silencing certain cancer promoting genes. This is done by complementing the cancer genes with the RNAi, such as keeping the mRNA sequences in accordance with the RNAi drug. Ideally, RNAi is should be injected and/or chemically modified so the RNAi can reach cancer cells more efficiently. RNAi uptake and regulation is controlled by the kidneys.

RNAi strategies also show potential for treating neurodegenerative diseases. Studies in cells and in mouse have shown that specifically targeting Amyloid beta-producing genes (e.g. BACE1 and APP) by RNAi can significantly reduced the amount of Aβ peptide which is correlated with the cause of Alzheimer's disease. In addition, this silencing-based approaches also provide promising results in treatment of Parkinson's disease and Polyglutamine disease.

The human immune system is divided into two separate branches: the innate immune system and the adaptive immune system. The innate immune system is the first defense against infection and responds to pathogens in a generic fashion. On the other hand, the adaptive immune system, a system that was evolved later than the innate, is composed mainly of highly specialized B and T cells that are trained to react to specific portions of pathogenic molecules.

The challenge between old pathogens and new has helped create a system of guarded cells and particles that are called safe framework. This framework has given humans an army of systems that search out and destroy invader particles, such as pathogens, microscopic organisms, parasites, and infections. The mammalian safe framework has developed to incorporate siRNA as a tool to indicate viral contamination, which has allowed siRNA is create an intense innate immune response.

siRNA is controlled by the innate immune system, which can be divided into the acute inflammatory responses and antiviral responses. The inflammatory response is created with signals from small signaling molecules, or cytokines. These include interleukin-1 (IL-1), interleukin-6 (IL-6), interleukin-12 (IL-12) and tumor necrosis factor α (TNF-α). The innate immune system generates inflammation and antiviral responses, which cause the release pattern recognition receptors (PRRs). These receptors help in labeling which pathogens are viruses, fungi, or bacteria. Moreover, the importance of siRNA and the innate immune system is to include more PRRs to help recognize different RNA structures. This makes it more likely for the siRNA to cause an immunostimulant response in the event of the pathogen.

RNAi has been used to genetically engineer plants to produce lower levels of natural plant toxins. Such techniques take advantage of the stable and heritable RNAi phenotype in plant stocks. Cotton seeds are rich in dietary protein but naturally contain the toxic terpenoid product gossypol, making them unsuitable for human consumption. RNAi has been used to produce cotton stocks whose seeds contain reduced levels of delta-cadinene synthase, a key enzyme in gossypol production without affecting the enzyme's production in other parts of the plant, where gossypol is itself important in preventing damage from plant pests.

Development efforts have successfully reduced the levels of allergens in tomato plants and fortification of plants such as tomatoes with dietary antioxidants. RNAi silencing of alpha-amylase have also been used to decrease Aspergillus flavus fungal growth in maize which would have otherwise contaminated the kernels with dangerous aflatoxins. Silencing lachrymatory factor synthase in onions have produced tearless onions and RNAi has been used in BP1 genes in rapeseeds to improve photosynthesis. SBEIIa and SBEIIb genes in wheat have been targeted in wheat in order to produce higher levels of amylose in order to improve bowel function, and Travella et al. 2006 employed RNAi for functional genomics an investigation of hexaploid bread races, while virus-induced gene silencing (VIGS, a subtype of RNAi) was used by Scofield et al. 2005 to investigate the mechanism of resistance provided by *Lr21* against wheat leaf rust in hexaploid wheat.

RNAi is under development as an insecticide, employing multiple approaches, including genetic engineering and topical application. Cells in the midgut of some insects take up the dsRNA molecules in the process referred to as environmental RNAi. In some insects the effect is systemic as the signal spreads throughout the insect's body (referred to as systemic RNAi).

Animals exposed to RNAi at doses millions of times higher than anticipated human exposure levels show no adverse effects. RNAi has varying effects in different species of Lepidoptera (butterflies and moths).

Drosophila spp., *Bombyx mori*, *Locusta* spp., *Spodoptera* spp., *Tribolium castaneum*, *Nilaparvata lugens*, *Helicoverpa armigera*, and *Apis mellifera* are models that have been widely used to learn how RNAi works within particular taxa of insects. *Musca domestica* has two *Ago2* genes and *Glossina morsitans* three, as found by Lewis et al. 2016 and Hain et al. 2010. In the case of the miRNA pathway *Diuraphis noxia* has two *Ago1*s, *M. domestica* two *Dcr1*s, *Acyrthosiphon pisum* two each of *Ago1* and *Lcqs* and *Dcr1* and four of *Pasha*. While in the piRNA, *G. morsitans* and *A. pisum* have two or three *Ago3*s each. This has led to identification of future insecticide development targets, and the modes of action and reasons for insecticide resistance of other insecticides.

Transgenic crops have been made to express dsRNA, carefully chosen to silence crucial genes in target pests. These dsRNAs are designed to affect only insects that express specific gene sequences. As a proof of principle, in 2009 a study showed RNAs that could kill any one of four fruit fly species while not harming the other three.

Alternatively dsRNA can be supplied without genetic engineering. One approach is to add them to irrigation water. The molecules are absorbed into the plants' vascular system and poison insects feeding on them. Another approach involves spraying dsRNA like a conventional pesticide. This would allow faster adaptation to resistance. Such approaches would require low cost sources of dsRNAs that do not currently exist.

Approaches to the design of genome-wide RNAi libraries can require more sophistication than the design of a single siRNA for a defined set of experimental conditions. Artificial neural networks are frequently used to design siRNA libraries and to predict their likely efficiency at gene knockdown. Mass genomic screening is widely seen as a promising method for genome annotation and has triggered the development of high-throughput screening methods based on microarrays.

Genome-scale RNAi research relies on high-throughput screening (HTS) technology. RNAi HTS technology allows genome-wide loss-of-function screening and is broadly used in the identification of genes associated with specific phenotypes. This technology has been hailed as a potential second genomics wave, following the first genomics wave of gene expression microarray and single nucleotide polymorphism discovery platforms. One major advantage of genome-scale RNAi screening is its ability to simultaneously interrogate thousands of genes. With the ability to generate a large amount of data per experiment, genome-scale RNAi screening has led to an explosion of data generation rates. Exploiting such large data sets is a fundamental challenge, requiring suitable statistics/bioinformatics methods. The basic process of cell-based RNAi screening includes the choice of an RNAi library, robust and stable cell types, transfection with RNAi agents, treatment/incubation, signal detection, analysis and identification of important genes or therapeutical targets.

CRISPR Screens

CRISPR-Cas9 has been used to delete genes in a multiplexed manner in cell-lines. Quantifying the amount of guide-RNAs for each gene before and after the experiment can point towards essential genes. If a guide-RNA disrupts an essential gene it will lead to the loss of that cell and hence there will be a depletion of that particular guide-RNA after the screen. In a recent CRISPR-cas9 experiment in mammalian cell-lines, around 2000 genes were found to be essential in multiple cell-lines. Some of these genes were essential in only one cell-line. Most of genes are part of multi-protein complexes. This approach can be used to identify synthetic lethality by using the appropriate genetic background. CRISPRi and CRISPRa enable loss-of-function and gain-of-function screens in a similar manner. CRISPRi identified ~2100 essential genes in the K562 cell-line. CRISPR deletion screens have also been used to identify potential regulatory elements of a gene. For example, a technique called ScanDel was published which attempted this approach. The authors deleted regions outside a gene of interest(HPRT1 involved in a Mendelian disorder) in an attempt to identify regulatory elements of this gene. Gassperini et al. did not identify any distal regulatory elements for HPRT1 using this approach, however such approaches can be extended to other genes of interest.

Functional Annotations for Genes

Genome Annotation

Putative genes can be identified by scanning a genome for regions likely to encode proteins, based on characteristics such as long open reading frames, transcriptional initiation sequences, and polyadenylation sites. A sequence identified as a putative gene must be confirmed by further evidence, such as similarity to cDNA or EST sequences from the same organism, similarity of the predicted protein sequence to known proteins, association with promoter sequences, or evidence that mutating the sequence produces an observable phenotype.

Rosetta Stone Approach

The Rosetta stone approach is a computational method for de-novo protein function prediction. It is based on the hypothesis that some proteins involved in a given physiological process may exist as two separate genes in one organism and as a single gene in another. Genomes are scanned for sequences that are independent in one organism and in a single open reading frame in another. If two genes have fused, it is predicted that they have similar biological functions that make such co-regulation advantageous.

14.1.3 Bioinformatics Methods for Functional Genomics

Because of the large quantity of data produced by these techniques and the desire to find biologically meaningful patterns, bioinformatics is crucial to analysis of functional genomics data. Examples of techniques in this class are data clustering or principal component analysis for unsupervised machine learning (class detection) as well as artificial neural networks or support vector machines for supervised machine learning (class prediction, classification). Functional enrichment analysis is used to determine the extent of over- or under-expression (positive- or negative-regulators in case of RNAi screens) of functional categories relative to a background sets. Gene ontology based enrichment analysis are provided by DAVID and gene set enrichment analysis (GSEA), pathway based analysis by Ingenuity and Pathway studio and protein complex based analysis by COMPLEAT.

New computational methods have been developed for understanding the results of a deep mutational scanning experiment. 'phydms' compares the result of a deep mutational scanning experiment to a phylogenetic tree. This allows the user to infer if the selection process in nature applies similar constraints on a protein as the results of the deep mutational scan indicate. This may allow an experimenter to choose between different experimental conditions based on how well they reflect nature. Deep mutational scanning has also been used to infer protein-protein interactions. The authors used a thermodynamic model to predict the effects of mutations in different parts of a dimer. Deep mutational structure can also be used to infer protein structure. Strong positive epistasis between two mutations in a deep mutational scan can be indicative of two parts of the protein that are close to each other in 3-D space. This information can then be used to infer protein structure. A proof of principle of this approach was shown by two groups using the protein GB1.

Results from MPRA experiments have required machine learning approaches to interpret the data. A gapped k-mer SVM model has been used to infer the kmers that are enriched within cis-regulatory sequences with high activity compared to sequences with lower activity. These models provide high predictive power. Deep learning and random forest approaches have also been used to interpret the results of these high-dimensional experiments. These models are beginning to help develop a better understanding of non-coding DNA function towards gene-regulation.

14.1.4 Consortium Projects

The ENCODE Project

The ENCODE (Encyclopedia of DNA elements) project is an in-depth analysis of the human genome whose goal is to identify all the functional elements of genomic DNA, in both coding and non-coding regions. Important results include evidence from genomic tiling arrays that most nucleotides are transcribed as coding transcripts, non-coding RNAs, or random transcripts, the discovery of additional transcriptional regulatory sites, further elucidation of chromatin-modifying mechanisms.

The Genotype-Tissue Expression (GTEx) Project

The GTEx project is a human genetics project aimed at understanding the role of genetic variation in shaping variation in the transcriptome across tissues. The project has collected a variety of tissue samples (> 50 different tissues) from more than 700 post-mortem donors. This has resulted in the collection of >11,000 samples. GTEx has helped understand the tissue-sharing and tissue-specificity of eQTLs. The genomic resource was developed to "enrich our understanding of how differences in our DNA sequence contribute to health and disease."

The Atlas of Variant Effects Alliance

The Atlas of Variant Effects Alliance (AVE), founded in 2020, is an international consortium aiming to catalog the impact of all possible genetic variants for disease-related functional genomics by creating variant effect maps that reveal the function of every possible single nucleotide change in a gene or regulatory element. AVE is funded in part through the Brotman Baty Institute at the University of Washington and the National Human Genome Research Institute, via funding from the Center of Excellence in Genome Science grant (NHGRI RM1HG010461).

14.2 CURRENT TRENDS IN FUNCTIONAL GENOMICS

14.2.1 High-Throughput Sequencing Technologies

Several new methods for DNA sequencing were developed in the mid to late 1990s and were implemented in commercial DNA sequencers by 2000. Together these were called the "next-generation" or "second-generation" sequencing (NGS) methods, in order to distinguish them from the earlier methods, including Sanger sequencing. In contrast to the first generation of sequencing, NGS technology is typically characterized by being highly scalable, allowing the entire genome to be sequenced at once. Usually, this is accomplished by fragmenting the genome into small pieces, randomly sampling for a fragment, and sequencing it using one of a variety of technologies, such as those described below. An entire genome is possible because multiple fragments are sequenced at once (giving it the name "massively parallel" sequencing) in an automated process.

NGS technology has tremendously empowered researchers to look for insights into health, anthropologists to investigate human origins, and is catalyzing the "Personalized Medicine" movement. However, it has also opened the door to more room for error. There are many software tools to carry out the computational analysis of NGS data, often compiled at online platforms such as CSI NGS Portal, each with its own algorithm. Even the parameters within one software package can change the outcome of the analysis. In addition, the large quantities of data produced by DNA sequencing have also required development of new methods and programs for sequence analysis. Several efforts to develop standards in the NGS field have been attempted to address these challenges, most of which have been small-scale efforts arising from individual labs. Most recently, a large, organized, FDA-funded effort has culminated in the BioCompute standard.

On 26 October 1990, Roger Tsien, Pepi Ross, Margaret Fahnestock and Allan J Johnston filed a patent describing stepwise ("base-by-base") sequencing with removable 3' blockers on DNA arrays (blots and single DNA molecules). In 1996, Pål Nyrén and his student Mostafa Ronaghi at the Royal Institute of Technology in Stockholm published their method of pyrosequencing.

On 1 April 1997, Pascal Mayer and Laurent Farinelli submitted patents to the World Intellectual Property Organization describing DNA colony sequencing. The DNA sample preparation and random surface-polymerase chain reaction (PCR) arraying methods described in this patent, coupled to Roger Tsien et al.'s "base-by-base" sequencing method, is now implemented in Illumina's Hi-Seq genome sequencers.

In 1998, Phil Green and Brent Ewing of the University of Washington described their phred quality score for sequencer data analysis, a landmark analysis technique that gained widespread adoption, and which is still the most common metric for assessing the accuracy of a sequencing platform.

Lynx Therapeutics published and marketed massively parallel signature sequencing (MPSS), in 2000. This method incorporated a parallelized, adapter/ligation-mediated, bead-based sequencing technology and served as the first commercially available "next-generation" sequencing method, though no DNA sequencers were sold to independent laboratories.

14.2.2 Transcriptomics and Gene Expression Profiling

Transcriptome

The transcriptome is the set of all RNA transcripts, including coding and non-coding, in an individual or a population of cells. The term can also sometimes be used to refer to all RNAs, or just mRNA, depending on the particular experiment. The term *transcriptome* is a portmanteau of the words *transcript* and *genome*; it is associated with the process of transcript production during the biological process of transcription.

The early stages of transcriptome annotations began with cDNA libraries published in the 1980s. Subsequently, the advent of high-throughput technology led to faster and more efficient ways of obtaining data about the transcriptome. Two biological techniques are used to study the transcriptome, namely DNA microarray, a hybridization-based technique and RNA-seq, a sequence-based approach. RNA-seq is the preferred method and has been the dominant transcriptomics technique since the 2010s. Single-cell transcriptomics allows tracking of transcript changes over time within individual cells.

Data obtained from the transcriptome is used in research to gain insight into processes such as cellular differentiation, carcinogenesis, transcription regulation and biomarker discovery among others. Transcriptome-obtained data also finds applications in establishing phylogenetic relationships during the process of evolution and in *in vitro* fertilization. The transcriptome is closely related to other -ome based biological fields of study; it is complementary to the proteome and the metabolome and encompasses the translatome, exome, meiome and thanatotranscriptome which can be seen as ome fields studying specific types of RNA transcripts. There are quantifiable and conserved relationships between the Transcriptome and other -omes, and Transcriptomics data can be used effectively to predict other molecular species, such as metabolites. There are numerous publicly available transcriptome databases.

Transcription

The transcriptome encompasses all the ribonucleic acid (RNA) transcripts present in a given organism or experimental sample. RNA is the main carrier of genetic information that is responsible for the process of converting DNA into an organism's phenotype. A gene can give rise to a single-stranded messenger RNA (mRNA) through a molecular process known as transcription; this mRNA is complementary to the strand of DNA it originated from. The enzyme RNA polymerase II attaches to the template DNA strand and catalyzes the addition of ribonucleotides to the 3' end of the growing sequence of the mRNA transcript.

In order to initiate its function, RNA polymerase II needs to recognize a promoter sequence, located upstream (5') of the gene. In eukaryotes, this process is mediated by transcription factors, most notably Transcription factor II D (TFIID) which recognizes the TATA box and aids in the positioning of RNA polymerase at the appropriate start site.

To finish the production of the RNA transcript, termination takes place usually several hundred nuclecotides away from the termination sequence and cleavage takes place. This process occurs in the nucleus of a cell along with RNA processing by which mRNA molecules are capped, spliced and polyadenylated to increase their stability before being subsequently taken to the cytoplasm. The mRNA gives rise to proteins through the process of translation that takes place in ribosomes.

Types of RNA Transcripts

Almost all functional transcripts are derived from known genes. The only exceptions are a small number of transcripts that might play a direct role in regulating gene expression near the prompters of known genes.

Gene occupy most of prokaryotic genomes so most of their genomes are transcribed. Many eukaryotic genomes are very large and known genes may take up only a fraction of the genome. In mammals, for example, known genes only account for 40-50% of the genome. Nevertheless, identified transcripts often map to a much larger fraction of the genome suggesting that the transcriptome contains spurious transcripts that do not come from genes. Some of these transcripts are known to be non-functional because they map to transcribed pseudogenes or degenerative transposons and viruses. Others map to unidentified regions of the genome that may be junk DNA.

Spurious transcription is very common in eukaryotes, especially those with large genomes that might contain a lot of junk DNA. Some scientists claim that if a transcript has not been assigned to a known gene then the default assumption must be that it is junk RNA until it has been shown to be functional. This would mean that much of the transcriptome in species with large genomes is probably junk RNA.

The transcriptome includes the transcripts of protein-coding genes (mRNA plus introns) as well as the transcripts of non-coding genes (functional RNAs plus introns).

- **Ribosomal RNA/rRNA:** Usually the most abundant RNA in the transcriptome.
- **Long Non-Coding RNA/lncRNA:** Non-coding RNA transcripts that are more than 200 nucleotides long. Members of this group comprise the largest fraction of the non-coding transcriptome other than introns. It is not known how many of these transcripts are functional and how many are junk RNA.
- Transfer RNA/tRNA
- **Micro RNA/miRNA:** 19-24 nucleotides (nt) long. Micro RNAs up- or downregulate expression levels of mRNAs by the process of RNA interference at the post-transcriptional level.

- **Small Interfering RNA/siRNA:** 20-24 nt
- Small Nucleolar RNA/snoRNA
- **Piwi-Interacting RNA/piRNA:** 24-31 nt. They interact with Piwi proteins of the Argonaute family and have a function in targeting and cleaving transposons.
- Enhancer RNA/eRNA:

Scope of Study

In the human genome, all genes get transcribed into RNA because that's how the molecular gene is defined. The transcriptome consists of coding regions of mRNA plus non-coding UTRs, introns, non-coding RNAs, and spurious non-functional transcripts.

Several factors render the content of the transcriptome difficult to establish. These include alternative splicing, RNA editing and alternative transcription among others. Additionally, transcriptome techniques are capable of capturing transcription occurring in a sample at a specific time point, although the content of the transcriptome can change during differentiation. The main aims of transcriptomics are the following: "catalogue all species of transcript, including mRNAs, non-coding RNAs and small RNAs; to determine the transcriptional structure of genes, in terms of their start sites, 5′ and 3′ ends, splicing patterns and other post-transcriptional modifications; and to quantify the changing expression levels of each transcript during development and under different conditions".

The term can be applied to the total set of transcripts in a given organism, or to the specific subset of transcripts present in a particular cell type. Unlike the genome which is roughly fixed for a given cell line (excluding mutations), the transcriptome can vary with external environmental conditions. Because it includes all mRNA transcripts in the cell, the transcriptome reflects the genes that are being actively expressed at any given time, with the exception of mRNA degradation phenomena such as transcriptional attenuation. The study of transcriptomics, (which includes expression profiling, splice variant analysis etc.), examines the expression level of RNAs in a given cell population, often focusing on mRNA, but sometimes including others such as tRNAs and sRNAs.

Methods of Construction

Transcriptomics is the quantitative science that encompasses the assignment of a list of strings ("reads") to the object ("transcripts" in the genome). To calculate the expression strength, the density of reads corresponding to each object is counted. Initially, transcriptomes were analyzed and studied using expressed sequence tags libraries and serial and cap analysis of gene expression (SAGE).

Currently, the two main transcriptomics techniques include DNA microarrays and RNA-Seq. Both techniques require RNA isolation through RNA extraction techniques, followed by its separation from other cellular components and enrichment of mRNA.

There are two general methods of inferring transcriptome sequences. One approach maps sequence reads onto a reference genome, either of the organism itself (whose transcriptome is being studied) or of a closely related species. The other approach, *de novo* transcriptome assembly, uses software to infer transcripts directly from short sequence reads and is used in organisms with genomes that are not sequenced.

DNA Microarrays

The first transcriptome studies were based on microarray techniques (also known as DNA chips). Microarrays consist of thin glass layers with spots on which oligonucleotides, known as "probes" are arrayed; each spot contains a known DNA sequence.

When performing microarray analyses, mRNA is collected from a control and an experimental sample, the latter usually representative of a disease. The RNA of interest is converted to cDNA to increase its stability and marked with fluorophores of two colors, usually green and red, for the two groups. The cDNA is spread onto the surface of the microarray where it hybridizes with oligonucleotides on the chip and a laser is used to scan. The fluorescence intensity on each spot of the microarray corresponds to the level of gene expression and based on the color of the fluorophores selected, it can be determined which of the samples exhibits higher levels of the mRNA of interest.

One microarray usually contains enough oligonucleotides to represent all known genes; however, data obtained using microarrays does not provide information about unknown genes. During the 2010s, microarrays were almost completely replaced by next-generation techniques that are based on DNA sequencing.

RNA Sequencing

RNA sequencing is a next-generation sequencing technology; as such it requires only a small amount of RNA and no previous knowledge of the genome. It allows for both qualitative and quantitative analysis of RNA transcripts, the former allowing discovery of new transcripts and the latter a measure of relative quantities for transcripts in a sample.

The three main steps of sequencing transcriptomes of any biological samples include RNA purification, the synthesis of an RNA or cDNA library and sequencing the library. The RNA purification process is different for short and long RNAs. This step is usually followed by an assessment of RNA quality, with the purpose of avoiding contaminants such as DNA or technical contaminants related to sample processing. RNA quality is measured using UV spectrometry with an absorbance peak of 260 nm. RNA integrity can also be analyzed quantitatively comparing the ratio and intensity of 28S RNA to 18S RNA reported in the RNA Integrity Number (RIN) score. Since mRNA is the species of interest and it represents only 3% of its total content, the RNA sample should be treated to remove rRNA and tRNA and tissue-specific RNA transcripts.

The step of library preparation with the aim of producing short cDNA fragments, begins with RNA fragmentation to transcripts in length between 50 and 300 base pairs. Fragmentation can be enzymatic (RNA endonucleases), chemical (trismagnesium salt buffer, chemical hydrolysis) or mechanical (sonication, nebulisation). Reverse transcription is used to convert the RNA templates into cDNA and three priming methods can be used to achieve it, including oligo-DT, using random primers or ligating special adaptor oligos.

Single-Cell Transcriptomics

Transcription can also be studied at the level of individual cells by single-cell transcriptomics. Single-cell RNA sequencing (scRNA-seq) is a recently developed technique that allows the analysis of the transcriptome of single cells, including bacteria. With single-cell transcriptomics, subpopulations of cell types that constitute the tissue of interest are also taken into consideration.

This approach allows to identify whether changes in experimental samples are due to phenotypic cellular changes as opposed to proliferation, with which a specific cell type might be overexpressed in the sample.

Additionally, when assessing cellular progression through differentiation, average expression profiles are only able to order cells by time rather than their stage of development and are consequently unable to show trends in gene expression levels specific to certain stages. Single-cell trarnscriptomic techniques have been used to characterize rare cell populations such as circulating tumor cells, cancer stem cells in solid tumors, and embryonic stem cells (ESCs) in mammalian blastocysts.

Although there are no standardized techniques for single-cell transcriptomics, several steps need to be undertaken. The first step includes cell isolation, which can be performed using low- and high-throughput techniques. This is followed by a qPCR step and then single-cell RNAseq where the RNA of interest is converted into cDNA. Newer developments in single-cell transcriptomics allow for tissue and sub-cellular localization preservation through cryo-sectioning thin slices of tissues and sequencing the transcriptome in each slice. Another technique allows the visualization of single transcripts under a microscope while preserving the spatial information of each individual cell where they are expressed.

Analysis

A number of organism-specific transcriptome databases have been constructed and annotated to aid in the identification of genes that are differentially expressed in distinct cell populations. RNA-seq is emerging (2013) as the method of choice for measuring transcriptomes of organisms, though the older technique of DNA microarrays is still used. RNA-seq measures the transcription of a specific gene by converting long RNAs into a library of cDNA fragments. The cDNA fragments are then sequenced using high-throughput sequencing technology and aligned to a reference genome or transcriptome which is then used to create an expression profile of the genes.

Applications

- **Mammals:**
  - The transcriptomes of stem cells and cancer cells are of particular interest to researchers who seek to understand the processes of cellular differentiation and carcinogenesis. A pipeline using RNA-seq or gene array data can be used to track genetic changes occurring in stem and precursor cells and requires at least three independent gene expression data from the former cell type and mature cells. Analysis of the transcriptomes of human oocytes and embryos is used to understand the molecular mechanisms and signaling pathways controlling early embryonic development, and could theoretically be a powerful tool in making proper embryo selection in in vitro fertilisation. Analyses of the transcriptome content of the placenta in the first-trimester of pregnancy in *in vitro* fertilization and embryo transfer (IVT-ET) revealed differences in genetic expression which are associated with higher frequency of adverse perinatal outcomes. Such insight can be used to optimize the practice. Transcriptome analyses can also be used to optimize cryopreservation of oocytes, by lowering injuries associated with the process.
  - Transcriptomics is an emerging and continually growing field in biomarker discovery for use in assessing the safety of drugs or chemical risk assessment. Transcriptomes may also be used to infer phylogenetic relationships among individuals or to detect evolutionary patterns of transcriptome conservation. Transcriptome analyses were used to discover the incidence of antisense transcription, their role in gene expression through interaction with surrounding genes and their abundance in different chromosomes. RNA-seq was also used to show how RNA isoforms, transcripts stemming from the same gene but with different structures, can produce complex phenotypes from limited genomes.
- **Plants:**
  - Transcriptome analysis have been used to study the evolution and diversification process of plant species. In 2014, the 1000 Plant Genomes Project

was completed in which the transcriptomes of 1,124 plant species from the families viridiplantae, glaucophyta and rhodophyta were sequenced. The protein coding sequences were subsequently compared to infer phylogenetic relationships between plants and to characterize the time of their diversification in the process of evolution. Transcriptome studies have been used to characterize and quantify gene expression in mature pollen. Genes involved in cell wall metabolism and cytoskeleton were found to be overexpressed. Transcriptome approaches also allowed to track changes in gene expression through different developmental stages of pollen, ranging from microspore to mature pollen grains; additionally such stages could be compared across species of different plants including *Arabidopsis*, rice and tobacco.

Relation to other Ome Fields

Similar to other -ome based technologies, analysis of the transcriptome allows for an unbiased approach when validating hypotheses experimentally. This approach also allows for the discovery of novel mediators in signaling pathways. As with other -omics based technologies, the transcriptome can be analyzed within the scope of a multiomics approach. It is complementary to metabolomics but contrary to proteomics, a direct association between a transcript and metabolite cannot be established.

There are several -ome fields that can be seen as subcategories of the transcriptome. The exome differs from the transcriptome in that it includes only those RNA molecules found in a specified cell population, and usually includes the amount or concentration of each RNA molecule in addition to the molecular identities. Additionally, the transcritpome also differs from the translatome, which is the set of RNAs undergoing translation.

The term meiome is used in functional genomics to describe the meiotic transcriptome or the set of RNA transcripts produced during the process of meiosis. Meiosis is a key feature of sexually reproducing eukaryotes, and involves the pairing of homologous chromosome, synapse and recombination. Since meiosis in most organisms occurs in a short time period, meiotic transcript profiling is difficult due to the challenge of isolation (or enrichment) of meiotic cells (meiocytes). As with transcriptome analyses, the meiome can be studied at a whole-genome level using large-scale transcriptomic techniques. The meiome has been well-characterized in mammal and yeast systems and somewhat less extensively characterized in plants.

The thanatotranscriptome consists of all RNA transcripts that continue to be expressed or that start getting re-expressed in internal organs of a dead body 24–48 hours following death. Some genes include those that are inhibited after fetal development. If the thanatotranscriptome is related to the process of programmed cell death (apoptosis), it can be referred to as the apoptotic thanatotranscriptome. Analyses of the thanatotranscriptome are used in forensic medicine.

eQTL mapping can be used to complement genomics with transcriptomics; genetic variants at DNA level and gene expression measures at RNA level.

Relation to Proteome

The transcriptome can be seen as a subset of the proteome, that is, the entire set of proteins expressed by a genome. However, the analysis of relative mRNA expression levels can be complicated by the fact that relatively small changes in mRNA expression can produce large changes in the total amount of the corresponding protein present in the cell. One analysis method, known as gene set enrichment analysis, identifies coregulated gene networks rather than individual genes that are up- or down-regulated in different cell populations.

Although microarray studies can reveal the relative amounts of different mRNAs in the cell, levels of mRNA are not directly proportional to the expression level of the proteins they code for. The number of protein molecules synthesized using a given mRNA molecule as a template is highly dependent on translation-initiation features of the mRNA sequence; in particular, the ability of the translation initiation sequence is a key determinant in the recruiting of ribosomes for protein translation.

Gene Expression Profiling

In the field of molecular biology, gene expression profiling is the measurement of the activity (the expression) of thousands of genes at once, to create a global picture of cellular function. These profiles can, for example, distinguish between cells that are actively dividing, or show how the cells react to a particular treatment. Many experiments of this sort measure an entire genome simultaneously, that is, every gene present in a particular cell. Several transcriptomics technologies can be used to generate the necessary data to analyse. DNA microarrays measure the relative activity of previously identified target genes. Sequence based techniques, like RNA-Seq, provide information on the sequences of genes in addition to their expression level.

Background

Expression profiling is a logical next step after sequencing a genome: the sequence tells us what the cell could possibly do, while the expression profile tells us what it is actually doing at a point in time. Genes contain the instructions for making messenger RNA (mRNA), but at any moment each cell makes mRNA from only a fraction of the genes it carries. If a gene is used to produce mRNA, it is considered "on", otherwise "off". Many factors determine whether a gene is on or off, such as the time of day, whether or not the cell is actively dividing, its local environment, and chemical signals from other cells. For instance, skin cells, liver cells and nerve cells turn on (express) somewhat different genes and that is in large part what makes them different. Therefore, an expression profile allows one to deduce a cell's type, state, environment, and so forth.

Expression profiling experiments often involve measuring the relative amount of mRNA expressed in two or more experimental conditions. This is because altered levels of a specific sequence of mRNA suggest a changed need for the protein coded by the mRNA, perhaps indicating a homeostatic response or a pathological condition. For example, higher levels of mRNA coding for alcohol dehydrogenase suggest that the cells or tissues under study are responding to increased levels of ethanol in their environment. Similarly, if breast cancer cells express higher levels of mRNA associated with a particular transmembrane receptor than normal cells do, it might be that this receptor plays a role in breast cancer. A drug that interferes with this receptor may prevent or treat breast cancer. In developing a drug, one may perform gene expression profiling experiments to help assess the drug's toxicity, perhaps by looking for changing levels in the expression of cytochrome P450 genes, which may be a biomarker of drug metabolism. Gene expression profiling may become an important diagnostic test.

Comparison to Proteomics

The human genome contains on the order of 20,000 genes which work in concert to produce roughly 1,000,000 distinct proteins. This is due to alternative splicing, and also because cells make important changes to proteins through posttranslational modification after they first construct them, so a given gene serves as the basis for many possible versions of a particular protein. In any case, a single mass spectrometry experiment can identify about 2,000 proteins or 0.2% of the total. While knowledge of the precise proteins a cell makes (proteomics) is more relevant than knowing how much messenger RNA is made from each gene, gene expression profiling provides the most global picture possible in a single experiment. However, proteomics methodology is improving. In other species, such as yeast, it is possible to identify over 4,000 proteins in just over one hour.

Use in Hypothesis Generation and Testing

Sometimes, a scientist already has an idea of what is going on, a hypothesis, and he or she performs an expression profiling experiment with the idea of potentially disproving this hypothesis. In other words, the scientist is making a specific prediction about levels of expression that could turn out to be false.

More commonly, expression profiling takes place before enough is known about how genes interact with experimental conditions for a testable hypothesis to exist. With no hypothesis, there is nothing to disprove, but expression profiling can help to identify a candidate hypothesis for future experiments. Most early expression profiling experiments, and many current ones, have this form which is known as class discovery. A popular approach to class discovery involves grouping similar genes or samples together using one of the many existing clustering methods such the traditional k-means or hierarchical clustering, or the more recent MCL. Apart from selecting a clustering algorithm, user usually has to choose an appropriate proximity measure (distance or similarity) between data objects.

Class prediction is more difficult than class discovery, but it allows one to answer questions of direct clinical significance such as, given this profile, what is the probability that this patient will respond to this drug? This requires many examples of profiles that responded and did not respond, as well as cross-validation techniques to discriminate between them.

Limitations

In general, expression profiling studies report those genes that showed statistically significant differences under changed experimental conditions. This is typically a small fraction of the genome for several reasons. First, different cells and tissues express a subset of genes as a direct consequence of cellular differentiation so many genes are turned off. Second, many of the genes code for proteins that are required for survival in very specific amounts so many genes do not change. Third, cells use many other mechanisms to regulate proteins in addition to altering the amount of mRNA, so these genes may stay consistently expressed even when protein concentrations are rising and falling. Fourth, financial constraints limit expression profiling experiments to a small number of observations of the same gene under identical conditions, reducing the statistical power of the experiment, making it impossible for the experiment to identify important but subtle changes. Finally, it takes a great amount of effort to discuss the biological significance of each regulated gene, so scientists often limit their discussion to a subset. Newer microarray analysis techniques automate certain aspects of attaching biological significance to expression profiling results, but this remains a very difficult problem.

The relatively short length of gene lists published from expression profiling experiments limits the extent to which experiments performed in different laboratories appear to agree. Placing expression profiling results in a publicly accessible microarray database makes it possible for researchers to assess expression patterns beyond the scope of published results, perhaps identifying similarity with their own work.

Validation of High Throughput Measurements

Both DNA microarrays and quantitative PCR exploit the preferential binding or "base pairing" of complementary nucleic acid sequences, and both are used in gene expression profiling, often in a serial fashion. While high throughput DNA microarrays lack the quantitative accuracy of qPCR, it takes about the same time to measure the gene expression of a few dozen genes via qPCR as it would to measure an entire genome using DNA microarrays. So it often makes sense to perform semi-quantitative DNA microarray analysis experiments to identify candidate genes, then perform qPCR on some of the most interesting candidate genes to validate the microarray results. Other experiments, such as a Western blot of some of the protein products of differentially expressed genes, make conclusions based on the expression profile more persuasive, since the mRNA levels do not necessarily correlate to the amount of expressed protein.

Statistical Analysis

Data analysis of microarrays has become an area of intense research. Simply stating that a group of genes were regulated by at least twofold, once a common practice, lacks a solid statistical footing. With five or fewer replicates in each group, typical for microarrays, a single outlier observation can create an apparent difference greater than two-fold. In addition, arbitrarily setting the bar at two-fold is not biologically sound, as it eliminates from consideration many genes with obvious biological significance.

Rather than identify differentially expressed genes using a fold change cutoff, one can use a variety of statistical tests or omnibus tests such as ANOVA, all of which consider both fold change and variability to create a p-value, an estimate of how often we would observe the data by chance alone. Applying p-values to microarrays is complicated by the large number of multiple comparisons (genes) involved. For example, a p-value of 0.05 is typically thought to indicate significance, since it estimates a 5% probability of observing the data by chance. But with 10,000 genes on a microarray, 500 genes would be identified as significant at $p < 0.05$ even if there were no difference between the experimental groups. One obvious solution is to consider significant only those genes meeting a much more stringent p value criterion, e.g., one could perform a Bonferroni correction on the p-values, or use a false discovery rate calculation to adjust p-values in proportion to the number of parallel tests involved. Unfortunately, these approaches may reduce the number of significant genes to zero, even when genes are in fact differentially expressed. Current statistics such as Rank products aim to strike a balance between false discovery of genes due to chance variation and non-discovery of differentially expressed genes. Commonly cited methods include the Significance Analysis of Microarrays (SAM) and a wide variety of methods are available from Bioconductor and a variety of analysis packages from bioinformatics companies.

Selecting a different test usually identifies a different list of significant genes since each test operates under a specific set of assumptions, and places a different emphasis on certain features in the data. Many tests begin with the assumption of a normal distribution in the data, because that seems like a sensible starting point and often produces results that appear more significant. Some tests consider the joint distribution of all gene observations to estimate general variability in measurements, while others look at each gene in isolation. Many modern microarray analysis techniques involve bootstrapping (statistics), machine learning or Monte Carlo methods.

As the number of replicate measurements in a microarray experiment increases, various statistical approaches yield increasingly similar results, but lack of concordance between different statistical methods makes array results appear less trustworthy. The MAQC Project makes recommendations to guide researchers in selecting more standard methods (e.g. using p-value and fold-change together for selecting the differentially expressed genes) so that experiments performed in different laboratories will agree better.

Different from the analysis on differentially expressed individual genes, another type of analysis focuses on differential expression or perturbation of pre-defined gene sets and is called gene set analysis. Gene set analysis demonstrated several major advantages over individual gene differential expression analysis. Gene sets are groups of genes that are functionally related according to current knowledge. Therefore, gene set analysis is considered a knowledge based analysis approach. Commonly used gene sets include those derived from KEGG pathways, Gene Ontology terms, gene groups that share some other functional annotations, such as common transcriptional regulators etc.

Representative gene set analysis methods include Gene Set Enrichment Analysis (GSEA), which estimates significance of gene sets based on permutation of sample labels, and Generally Applicable Gene-set Enrichment (GAGE), which tests the significance of gene sets based on permutation of gene labels or a parametric distribution.

Gene Annotation

While the statistics may identify which gene products change under experimental conditions, making biological sense of expression profiling rests on knowing which protein each gene product makes and what function this protein performs. Gene annotation provides functional and other information, for example the location of each gene within a particular chromosome.

Some functional annotations are more reliable than others; some are absent. Gene annotation databases change regularly, and various databases refer to the same protein by different names, reflecting a changing understanding of protein function. Use of standardized gene nomenclature helps address the naming aspect of the problem, but exact matching of transcripts to genes remains an important consideration.

Categorizing Regulated Genes

Having identified some set of regulated genes, the next step in expression profiling involves looking for patterns within the regulated set. Do the proteins made from these genes perform similar functions? Are they chemically similar? Do they reside in similar parts of the cell? Gene ontology analysis provides a standard way to define these relationships. Gene ontologies start with very broad categories, e.g., "metabolic process" and break them down into smaller categories, e.g., "carbohydrate metabolic process" and finally into quite restrictive categories like "inositol and derivative phosphorylation".

Genes have other attributes beside biological function, chemical properties and cellular location. One can compose sets of genes based on proximity to other genes, association with a disease, and relationships with drugs or toxins. The Molecular Signatures Database and the Comparative Toxicogenomics Database are examples of resources to categorize genes in numerous ways.

Finding Patterns Among Regulated Genes

Regulated genes are categorized in terms of what they are and what they do, important relationships between genes may emerge. For example, we might see evidence that a certain gene creates a protein to make an enzyme that activates a protein to turn on a second gene on our list. This second gene may be a transcription factor that regulates yet another gene from our list. Observing these links we may begin to suspect that they represent much more than chance associations in the results, and that they are all on our list because of an underlying biological process. On the other hand, it could be that if one selected genes at random, one might find many that seem to have something in common. In this sense, we need rigorous statistical procedures to test whether the emerging biological themes is significant or not. That is where gene set analysis comes in.

Cause and Effect Relationships

Fairly straightforward statistics provide estimates of whether associations between genes on lists are greater than what one would expect by chance. These statistics are interesting, even if they represent a substantial oversimplification of what is really going on. Here is an example. Suppose there are 10,000 genes in an experiment, only 50 (0.5%) of which play a known role in making cholesterol. The experiment identifies 200 regulated genes. Of those, 40 (20%) turn out to be on a list of cholesterol genes as well. Based on the overall prevalence of the cholesterol genes (0.5%) one expects an average of 1 cholesterol gene for every 200 regulated genes, that is, 0.005 times 200. This expectation is an average, so one expects to see more than one some of the time. The question becomes how often we would see 40 instead of 1 due to pure chance.

According to the hypergeometric distribution, one would expect to try about 10^57 times (10 followed by 56 zeroes) before picking 39 or more of the cholesterol genes from a pool of 10,000 by drawing 200 genes at random. Whether one pays much attention to how infinitesimally small the probability of observing this by chance is, one would conclude that the regulated gene list is enriched in genes with a known cholesterol association.

One might further hypothesize that the experimental treatment regulates cholesterol, because the treatment seems to selectively regulate genes associated with cholesterol. While this may be true, there are a number of reasons why making this a firm conclusion based on enrichment alone represents an unwarranted leap of faith. One previously mentioned issue has to do with the observation that gene regulation may have no direct impact on protein regulation: even if the proteins coded for by these genes do nothing other than make cholesterol, showing that their mRNA is altered does not directly tell us what is happening at the protein level. It is quite possible that the amount of these cholesterol-related proteins remains constant under the experimental conditions. Second, even if protein levels do change, perhaps there is always enough of them around to make cholesterol as fast as it can be possibly made, that is, another protein, not on our list, is the rate determining step in the process of making cholesterol. Finally, proteins typically play many roles, so these genes may be regulated not because of their shared association with making cholesterol but because of a shared role in a completely independent process.

Bearing the foregoing caveats in mind, while gene profiles do not in themselves prove causal relationships between treatments and biological effects, they do offer unique biological insights that would often be very difficult to arrive at in other ways.

Using Patterns to Find Regulated Genes

As described above, one can identify significantly regulated genes first and then find patterns by comparing the list of significant genes to sets of genes known to share certain associations. One can also work the problem in reverse order. Here is a very simple example. Suppose there are 40 genes associated with a known process, for example, a predisposition to diabetes. Looking at two groups of expression profiles, one for mice fed a high carbohydrate diet and one for mice fed a low carbohydrate diet, one observes that all 40 diabetes genes are expressed at a higher level in the high carbohydrate group than the low carbohydrate group. Regardless of whether any of these genes would have made it to a list of significantly altered genes, observing all 40 up, and none down appears unlikely to be the result of pure chance: flipping 40 heads in a row is predicted to occur about one time in a trillion attempts using a fair coin.

For a type of cell, the group of genes whose combined expression pattern is uniquely characteristic to a given condition constitutes the gene signature of this condition. Ideally, the gene signature can be used to select a group of patients at a specific state of a disease with accuracy that facilitates selection of treatments. Gene Set Enrichment Analysis (GSEA) and similar methods take advantage of this kind of logic but uses more sophisticated statistics, because component genes in real processes display more complex behavior than simply moving up or down as a group, and the amount the genes move up and down is meaningful, not just the direction. In any case, these statistics measure how different the behavior of some small set of genes is compared to genes not in that small set.

GSEA uses a Kolmogorov Smirnov style statistic to see whether any previously defined gene sets exhibited unusual behavior in the current expression profile. This leads to a multiple hypothesis testing challenge, but reasonable methods exist to address it.

14.2.3 Proteomics and Metabolomics

Proteomics

Proteomics is the large-scale study of proteins. Proteins are vital macromolecules of all living organisms, with many functions such as the formation of structural fibers of muscle tissue, enzymatic digestion of food, or synthesis and replication of DNA. In addition, other kinds of proteins include antibodies that protect an organism from infection, and hormones that send important signals throughout the body. The proteome is the entire set of proteins produced or modified by an organism or system. Proteomics enables the identification of ever-increasing numbers of proteins. This varies with time and distinct requirements, or stresses, that a cell or organism undergoes.

Proteomics is an interdisciplinary domain that has benefited greatly from the genetic information of various genome projects, including the Human Genome Project. It covers the exploration of proteomes from the overall level of protein composition, structure, and activity, and is an important component of functional genomics. *Proteomics* generally denotes the large-scale experimental analysis of proteins and proteomes, but often refers specifically to protein purification and mass spectrometry. Indeed, mass spectrometry is the most powerful method for analysis of proteomes, both in large samples composed of millions of cells and in single cells.

Complexity of the Problem

After genomics and transcriptomics, proteomics is the next step in the study of biological systems. It is more complicated than genomics because an organism's genome is more or less constant, whereas proteomes differ from cell to cell and from time to time. Distinct genes are expressed in different cell types, which means that even the basic set of proteins produced in a cell must be identified.

In the past this phenomenon was assessed by RNA analysis, which was found to lack correlation with protein content. It is now known that mRNA is not always translated into protein, and the amount of protein produced for a given amount of mRNA depends on the gene it is transcribed from and on the cell's physiological state. Proteomics confirms the presence of the protein and provides a direct measure of its quantity.

Post-Translational Modifications

Not only does the translation from mRNA cause differences, but many proteins also are subjected to a wide variety of chemical modifications after translation. The most common and widely studied post-translational modifications include phosphorylation and glycosylation. Many of these post-translational modifications are critical to the protein's function.

- **Phosphorylation:** One such modification is phosphorylation, which happens to many enzymes and structural proteins in the process of cell signaling. The addition of a phosphate to particular amino acids—most commonly serine and threonine mediated by serine-threonine kinases, or more rarely tyrosine mediated by tyrosine kinases—causes a protein to become a target for binding or interacting with a distinct set of other proteins that recognize the phosphorylated domain. Because protein phosphorylation is one of the most studied protein modifications, many "proteomic" efforts are geared to determining the set of phosphorylated proteins in a particular cell or tissue-type under particular circumstances. This alerts the scientist to the signaling pathways that may be active in that instance.
- **Ubiquitination:** Ubiquitin is a small protein that may be affixed to certain protein substrates by enzymes called E3 ubiquitin ligases. Determining which proteins are poly-ubiquitinated helps understand how protein pathways are regulated. This is, therefore, an additional legitimate "proteomic" study. Similarly, once a researcher determines which substrates are ubiquitinated by each ligase, determining the set of ligases expressed in a particular cell type is helpful.
- **Additional Modifications:** In addition to phosphorylation and ubiquitination, proteins may be subjected to (among others) methylation, acetylation, glycosylation, oxidation, and nitrosylation. Some proteins undergo all these modifications, often in time-dependent combinations. This illustrates the potential complexity of studying protein structure and function.

Distinct Proteins are Made under Distinct Settings

A cell may make different sets of proteins at different times or under different conditions, for example during development, cellular differentiation, cell cycle, or carcinogenesis. Further increasing proteome complexity, as mentioned, most proteins are able to undergo a wide range of post-translational modifications.

Therefore, a "proteomics" study may become complex very quickly, even if the topic of study is restricted. In more ambitious settings, such as when a biomarker for a specific cancer subtype is sought, the proteomics scientist might elect to study multiple blood serum samples from multiple cancer patients to minimise confounding factors and account for experimental noise. Thus, complicated experimental designs are sometimes necessary to account for the dynamic complexity of the proteome.

Limitations of Genomics and Proteomics Studies

Proteomics gives a different level of understanding than genomics for many reasons:

- The level of transcription of a gene gives only a rough estimate of its *level of translation* into a protein. An mRNA produced in abundance may be degraded rapidly or translated inefficiently, resulting in a small amount of protein.
- Many proteins experience *post-translational modifications* that profoundly affect their activities; for example, some proteins are not active until they become phosphorylated. Methods such as phosphoproteomics and glycoproteomics are used to study post-translational modifications.
- Many transcripts give rise to more than one protein, through alternative splicing or alternative post-translational modifications.
- Many proteins form complexes with other proteins or RNA molecules, and only function in the presence of these other molecules.
- Protein degradation rate plays an important role in protein content.

Reproducibility

One major factor affecting reproducibility in proteomics experiments is the simultaneous elution of many more peptides than mass spectrometers can measure. This causes stochastic differences between experiments due to data-dependent acquisition of tryptic peptides. Although early large-scale shotgun proteomics analyses showed considerable variability between laboratories, presumably due in part to technical and experimental differences between laboratories, reproducibility has been improved in more recent mass spectrometry analysis, particularly on the protein level. Notably, targeted proteomics shows increased reproducibility and repeatability compared with shotgun methods, although at the expense of data density and effectiveness.

Data Quality

Proteomic analysis is highly amenable to automation and large data sets are created, which are processed by software algorithms. Filter parameters are used to reduce the number of false hits, but they cannot be completely eliminated. Scientists have expressed the need for awareness that proteomics experiments should adhere to the criteria of analytical chemistry (sufficient data quality, sanity check, validation).

Methods of Studying Proteins

In proteomics, there are multiple methods to study proteins. Generally, proteins may be detected by using either antibodies (immunoassays), electrophoretic separation or mass spectrometry. If a complex biological sample is analyzed, either a very specific antibody needs to be used in quantitative dot blot analysis (QDB), or biochemical separation then needs to be used before the detection step, as there are too many analytes in the sample to perform accurate detection and quantification.

Protein Detection with Antibodies (Immunoassays)

Antibodies to particular proteins, or their modified forms, have been used in biochemistry and cell biology studies. These are among the most common tools used by molecular biologists today. There are several specific techniques and protocols that use antibodies for protein detection. The enzyme-linked immunosorbent assay (ELISA) has been used for decades to detect and quantitatively measure proteins in samples. The western blot may be used for detection and quantification of individual proteins, where in an initial step, a complex protein mixture is separated using SDS-PAGE and then the protein of interest is identified using an antibody.

Modified proteins may be studied by developing an antibody specific to that modification. For example, some antibodies only recognize certain proteins when they are tyrosine-phosphorylated, they are known as phospho-specific antibodies. Also, there are antibodies specific to other modifications. These may be used to determine the set of proteins that have undergone the modification of interest.

Immunoassays can also be carried out using recombinantly generated immunoglobulin derivatives or synthetically designed protein scaffolds that are selected for high antigen specificity. Such binders include single domain antibody fragments (Nanobodies), designed ankyrin repeat proteins (DARPins) and aptamers.

Disease detection at the molecular level is driving the emerging revolution of early diagnosis and treatment. A challenge facing the field is that protein biomarkers for early diagnosis may be present in very low abundance. The lower limit of detection with conventional immunoassay technology is the upper femtomolar range (10^{-13} M). Digital immunoassay technology has improved detection sensitivity three logs, to the attomolar range (10^{-16} M). This capability has the potential to open new advances in diagnostics and therapeutics, but such technologies have been relegated to manual procedures that are not well suited for efficient routine use.

Antibody-Free Protein Detection

While protein detection with antibodies is still very common in molecular biology, other methods have been developed as well, that do not rely on an antibody. These methods offer various advantages, for instance they often are able to determine the sequence of a protein or peptide, they may have higher throughput than antibody-based, and they sometimes can identify and quantify proteins for which no antibody exists.

- **Detection Methods:** One of the earliest methods for protein analysis has been Edman degradation (introduced in 1967) where a single peptide is subjected to multiple steps of chemical degradation to resolve its sequence. These early methods have mostly been supplanted by technologies that offer higher throughput. More recently implemented methods use mass spectrometry-based techniques, a development that was made possible by the discovery of "soft ionization" methods developed in the 1980s, such as matrix-assisted laser desorption/ ionization (MALDI) and electrospray ionization (ESI). These methods gave rise to the top-down and the bottom-up proteomics workflows where often additional separation is performed before analysis.
- **Separation Methods:** For the analysis of complex biological samples, a reduction of sample complexity is required. This may be performed off-line by one-dimensional or two-dimensional separation. More recently, on-line methods have been developed where individual peptides (in bottom-up proteomics approaches) are separated using reversed-phase chromatography and then, directly ionized using ESI; the direct coupling of separation and analysis explains the term "on-line" analysis.

Hybrid Technologies

Several hybrid technologies use antibody-based purification of individual analytes and then perform mass spectrometric analysis for identification and quantification. Examples of these methods are the MSIA (mass spectrometric immunoassay), developed by Randall Nelson in 1995, and the SISCAPA (Stable Isotope Standard Capture with Anti-Peptide Antibodies) method, introduced by Leigh Anderson in 2004.

Current Research Methodologies

Fluorescence two-dimensional differential gel electrophoresis (2-D DIGE) may be used to quantify variation in the 2-D DIGE process and establish statistically valid thresholds for assigning quantitative changes between samples.

Comparative proteomic analysis may reveal the role of proteins in complex biological systems, including reproduction. For example, treatment with the insecticide triazophos causes an increase in the content of brown planthopper (*Nilaparvata lugens* (Stål)) male accessory gland proteins (Acps) that may be transferred to females via mating, causing an increase in fecundity (i.e. birth rate) of females. To identify changes in the types of accessory gland proteins (Acps) and reproductive proteins that mated female planthoppers received from male planthoppers, researchers conducted a comparative proteomic analysis of mated *N. lugens* females. The results indicated that these proteins participate in the reproductive process of *N. lugens* adult females and males.

Proteome analysis of *Arabidopsis peroxisomes* has been established as the major unbiased approach for identifying new peroxisomal proteins on a large scale.

There are many approaches to characterizing the human proteome, which is estimated to contain between 20,000 and 25,000 non-redundant proteins. The number of unique protein species likely will increase by between 50,000 and 500,000 due to RNA splicing and proteolysis events, and when post-translational modification also are considered, the total number of unique human proteins is estimated to range in the low millions.

In addition, the first promising attempts to decipher the proteome of animal tumors have recently been reported. This method was used as a functional method in *Macrobrachium rosenbergii* protein profiling.

High-Throughput Proteomic Technologies

Proteomics has steadily gained momentum over the past decade with the evolution of several approaches. Few of these are new, and others build on traditional methods. Mass spectrometry-based methods, affinity proteomics, and micro arrays are the most common technologies for large-scale study of proteins.

Mass Spectrometry and Protein Profiling

There are two mass spectrometry-based methods currently used for protein profiling. The more established and widespread method uses high resolution, two-dimensional electrophoresis to separate proteins from different samples in parallel, followed by selection and staining of differentially expressed proteins to be identified by mass spectrometry. Despite the advances in 2-DE and its maturity, it has its limits as well. The central concern is the inability to resolve all the proteins within a sample, given their dramatic range in expression level and differing properties. The combination of pore size, and protein charge, size and shape can greatly determine migration rate which leads to other complications.

The second quantitative approach uses stable isotope tags to differentially label proteins from two different complex mixtures. Here, the proteins within a complex mixture are labeled isotopically first, and then digested to yield labeled peptides. The labeled mixtures are then combined, the peptides separated by multidimensional liquid chromatography and analyzed by tandem mass spectrometry. Isotope coded affinity tag (ICAT) reagents are the widely used isotope tags. In this method, the cysteine residues of proteins get covalently attached to the ICAT reagent, thereby reducing the complexity of the mixtures omitting the non-cysteine residues.

Quantitative proteomics using stable isotopic tagging is an increasingly useful tool in modern development.

Firstly, chemical reactions have been used to introduce tags into specific sites or proteins for the purpose of probing specific protein functionalities. The isolation of phosphorylated peptides has been achieved using isotopic labeling and selective chemistries to capture the fraction of protein among the complex mixture.

Secondly, the ICAT technology was used to differentiate between partially purified or purified macromolecular complexes such as large RNA polymerase II pre-initiation complex and the proteins complexed with yeast transcription factor. Thirdly, ICAT labeling was recently combined with chromatin isolation to identify and quantify chromatin-associated proteins. Finally ICAT reagents are useful for proteomic profiling of cellular organelles and specific cellular fractions.

Another quantitative approach is the accurate mass and time (AMT) tag approach developed by Richard D. Smith and coworkers at Pacific Northwest National Laboratory.

In this approach, increased throughput and sensitivity is achieved by avoiding the need for tandem mass spectrometry, and making use of precisely determined separation time information and highly accurate mass determinations for peptide and protein identifications.

Affinity Proteomics

Affinity proteomics uses antibodies or other affinity reagents (such as oligonucleotide-based aptamers) as protein-specific detection probes. Currently this method can interrogate several thousand proteins, typically from biofluids such as plasma, serum or cerebrospinal fluid (CSF). A key differentiator for this technology is the ability to analyze hundreds or thousands of samples in a reasonable timeframe (a matter of days or weeks); mass spectrometry-based methods are not scalable to this level of sample throughput for proteomics analyses.

Protein Chips

Balancing the use of mass spectrometers in proteomics and in medicine is the use of protein micro arrays. The aim behind protein micro arrays is to print thousands of protein detecting features for the interrogation of biological samples. Antibody arrays are an example in which a host of different antibodies are arrayed to detect their respective antigens from a sample of human blood. Another approach is the arraying of multiple protein types for the study of properties like protein-DNA, protein-protein and protein-ligand interactions. Ideally, the functional proteomic arrays would contain the entire complement of the proteins of a given organism. The first version of such arrays consisted of 5000 purified proteins from yeast deposited onto glass microscopic slides. Despite the success of first chip, it was a greater challenge for protein arrays to be implemented. Proteins are inherently much more difficult to work with than DNA. They have a broad dynamic range, are less stable than DNA and their structure is difficult to preserve on glass slides, though they are essential for most assays. The global ICAT technology has striking advantages over protein chip technologies.

Reverse-Phased Protein Microarrays

This is a promising and newer microarray application for the diagnosis, study and treatment of complex diseases such as cancer. The technology merges laser capture microdissection (LCM) with micro array technology, to produce reverse-phase protein microarrays. In this type of microarrays, the whole collection of protein themselves are immobilized with the intent of capturing various stages of disease within an individual patient. When used with LCM, reverse phase arrays can monitor the fluctuating state of proteome among different cell population within a small area of human tissue. This is useful for profiling the status of cellular signaling molecules, among a cross-section of tissue that includes both normal and cancerous cells. This approach is useful in monitoring the status of key factors in normal prostate epithelium and invasive prostate cancer tissues. LCM then dissects these tissue and protein lysates were arrayed onto nitrocellulose slides, which were probed with specific antibodies. This method can track all kinds of molecular events and can compare diseased and healthy tissues within the same patient enabling the development of treatment strategies and diagnosis. The ability to acquire proteomics snapshots of neighboring cell populations, using reverse-phase microarrays in conjunction with LCM has a number of applications beyond the study of tumors. The approach can provide insights into normal physiology and pathology of all the tissues and is invaluable for characterizing developmental processes and anomalies.

Protein Detection via Bioorthogonal Chemistry

Recent advancements in bioorthogonal chemistry have revealed applications in protein analysis. The extension of using organic molecules to observe their reaction with proteins reveals extensive methods to tag them. Unnatural amino acids and various functional groups represent new growing technologies in proteomics.

Specific biomolecules that are capable of being metabolized in cells or tissues are inserted into proteins or glycans. The molecule will have an affinity tag, modifying the protein allowing it to be detected. Azidohomoalanine (AHA) utilizes this affinity tag via incorporation with Met-t-RNA synthetase to incorporate into proteins. This has allowed AHA to assist in determine the identity of newly synthesized proteins created in response to perturbations and to identify proteins secreted by cells.

Recent studies using ketones and aldehydes condensations show that they are best suited for in vitro or cell surface labeling. However, using ketones and aldehydes as bioorthogonal reporters revealed slow kinetics indicating that while effective for labeling, the concentration must be high.

Certain proteins can be detected via their reactivity to azide groups. Non-proteinogenic amino acids can bear azide groups which react with phosphines in Staudinger ligations. This reaction has already been used to label other biomolecules in living cells and animals.

The bioorthoganal field is expanding and is driving further applications within proteomics. It is worthwhile noting the limitations and benefits. Rapid reactions can create bioconjuctions and create high concentrations with low amounts of reactants. Contrarily slow kinetic reactions like aldehyde and ketone condensation while effective require a high concentration making it cost inefficient.

Practical Applications

- **New Drug Discovery:** One major development to come from the study of human genes and proteins has been the identification of potential new drugs for the treatment of disease. This relies on genome and proteome information to identify proteins associated with a disease, which computer software can then use as targets for new drugs. For example, if a certain protein is implicated in a disease, its 3D structure provides the information to design drugs to interfere with the action of the protein. A molecule that fits the active site of an enzyme, but cannot be released by the enzyme, inactivates the enzyme. This is the basis of new drug-discovery tools, which aim to find new drugs to inactivate proteins involved in disease. As genetic differences among individuals are found, researchers expect to use these techniques to develop personalized drugs that are more effective for the individual. Proteomics is also used to reveal complex plant-insect interactions that help identify candidate genes involved in the defensive response of plants to herbivory. A branch of proteomics called chemoproteomics provides numerous tools and techniques to detect protein targets of drugs.

- **Interaction Proteomics and Protein Networks:** Interaction proteomics is the analysis of protein interactions from scales of binary interactions to proteome- or network-wide. Most proteins function via protein–protein interactions, and one goal of interaction proteomics is to identify binary protein interactions, protein complexes, and interactomes. Several methods are available to probe protein–protein interactions. While the most traditional method is yeast two-hybrid analysis, a powerful emerging method is affinity purification followed by protein mass spectrometry using tagged protein baits. Other methods include surface plasmon resonance (SPR), protein microarrays, dual polarisation interferometry, microscale thermophoresis, kinetic exclusion assay, and experimental methods such as phage display and *in silico* computational methods. Knowledge of protein-protein interactions is especially useful in regard to biological networks and systems biology, for example in cell signaling cascades and gene regulatory networks (GRNs, where knowledge of protein-DNA interactions is also informative). Proteome-wide analysis of protein interactions, and integration of these interaction patterns into larger biological networks, is crucial towards understanding systems-level biology.
- **Expression Proteomics:** Expression proteomics includes the analysis of protein expression at a larger scale. It helps identify main proteins in a particular sample, and those proteins differentially expressed in related samples—such as diseased vs. healthy tissue. If a protein is found only in a diseased sample then it can be a useful drug target or diagnostic marker. Proteins with the same or similar expression profiles may also be functionally related. There are technologies such as 2D-PAGE and mass spectrometry that are used in expression proteomics.
- **Biomarkers:** The National Institutes of Health has defined a biomarker as "a characteristic that is objectively measured and evaluated as an indicator of normal biological processes, pathogenic processes, or pharmacologic responses to a therapeutic intervention." Understanding the proteome, the structure and function of each protein and the complexities of protein–protein interactions are critical for developing the most effective diagnostic techniques and disease treatments in the future. For example, proteomics is highly useful in the identification

of candidate biomarkers (proteins in body fluids that are of value for diagnosis), identification of the bacterial antigens that are targeted by the immune response, and identification of possible immunohistochemistry markers of infectious or neoplastic diseases. An interesting use of proteomics is using specific protein biomarkers to diagnose disease. A number of techniques allow to test for proteins produced during a particular disease, which helps to diagnose the disease quickly. Techniques include western blot, immunohistochemical staining, enzyme linked immunosorbent assay (ELISA) or mass spectrometry. Secretomics, a subfield of proteomics that studies secreted proteins and secretion pathways using proteomic approaches, has recently emerged as an important tool for the discovery of biomarkers of disease.

- **Proteogenomics:** In proteogenomics, proteomic technologies such as mass spectrometry are used for improving gene annotations. Parallel analysis of the genome and the proteome facilitates discovery of post-translational modifications and proteolytic events, especially when comparing multiple species (comparative proteogenomics).
- **Structural Proteomics:** Structural proteomics includes the analysis of protein structures at large-scale. It compares protein structures and helps identify functions of newly discovered genes. The structural analysis also helps to understand that where drugs bind to proteins and also shows where proteins interact with each other. This understanding is achieved using different technologies such as X-ray crystallography and NMR spectroscopy.

Bioinformatics for Proteomics (Proteome Informatics)

Much proteomics data is collected with the help of high throughput technologies such as mass spectrometry and microarray. It would often take weeks or months to analyze the data and perform comparisons by hand. For this reason, biologists and chemists are collaborating with computer scientists and mathematicians to create programs and pipeline to computationally analyze the protein data. Using bioinformatics techniques, researchers are capable of faster analysis and data storage. A good place to find lists of current programs and databases is on the ExPASy bioinformatics resource portal. The applications of bioinformatics-based proteomics include medicine, disease diagnosis, biomarker identification, and many more.

- **Protein Identification:** Mass spectrometry and microarray produce peptide fragmentation information but do not give identification of specific proteins present in the original sample. Due to the lack of specific protein identification, past researchers were forced to decipher the peptide fragments themselves. However, there are currently programs available for protein identification. These programs take the peptide sequences output from mass spectrometry and microarray and return information about matching or similar proteins. This is done through algorithms implemented by the program which perform alignments with proteins from known databases such as UniProt and PROSITE to predict what proteins are in the sample with a degree of certainty.

- **Protein Structure:** The biomolecular structure forms the 3D configuration of the protein. Understanding the protein's structure aids in the identification of the protein's interactions and function. It used to be that the 3D structure of proteins could only be determined using X-ray crystallography and NMR spectroscopy. As of 2017, Cryo-electron microscopy is a leading technique, solving difficulties with crystallization (in X-ray crystallography) and conformational ambiguity (in NMR); resolution was 2.2Å as of 2015. Now, through bioinformatics, there are computer programs that can in some cases predict and model the structure of proteins. These programs use the chemical properties of amino acids and structural properties of known proteins to predict the 3D model of sample proteins. This also allows scientists to model protein interactions on a larger scale. In addition, biomedical engineers are developing methods to factor in the flexibility of protein structures to make comparisons and predictions.
- **Post-Translational Modifications:** Most programs available for protein analysis are not written for proteins that have undergone post-translational modifications. Some programs will accept post-translational modifications to aid in protein identification but then ignore the modification during further protein analysis. It is important to account for these modifications since they can affect the protein's structure. In turn, computational analysis of post-translational modifications has gained the attention of the scientific community. The current post-translational modification programs are only predictive. Chemists, biologists and computer scientists are working together to create and introduce new pipelines that allow for analysis of post-translational modifications that have been experimentally identified for their effect on the protein's structure and function.
- **Computational Methods in Studying Protein Biomarkers:** One example of the use of bioinformatics and the use of computational methods is the study of protein biomarkers. Computational predictive models have shown that extensive and diverse feto-maternal protein

trafficking occurs during pregnancy and can be readily detected non-invasively in maternal whole blood. This computational approach circumvented a major limitation, the abundance of maternal proteins interfering with the detection of fetal proteins, to fetal proteomic analysis of maternal blood. Computational models can use fetal gene transcripts previously identified in maternal whole blood to create a comprehensive proteomic network of the term neonate. Such work shows that the fetal proteins detected in pregnant woman's blood originate from a diverse group of tissues and organs from the developing fetus. The proteomic networks contain many biomarkers that are proxies for development and illustrate the potential clinical application of this technology as a way to monitor normal and abnormal fetal development. An information-theoretic framework has also been introduced for biomarker discovery, integrating biofluid and tissue information. This new approach takes advantage of functional synergy between certain biofluids and tissues with the potential for clinically significant findings not possible if tissues and biofluids were considered individually. By conceptualizing tissue-biofluid as information channels, significant biofluid proxies can be identified and then used for the guided development of clinical diagnostics. Candidate biomarkers are then predicted based on information transfer criteria across the tissue-biofluid channels. Significant biofluid-tissue relationships can be used to prioritize clinical validation of biomarkers.

Emerging Trends

A number of emerging concepts have the potential to improve the current features of proteomics. Obtaining absolute quantification of proteins and monitoring post-translational modifications are the two tasks that impact the understanding of protein function in healthy and diseased cells. Further, the throughput and sensitivity of proteomic assays, often measured as samples analyzed per day and depth of proteome coverage, respectively, have driven development of cutting-edge instrumentation and methodologies.

For many cellular events, the protein concentrations do not change; rather, their function is modulated by post-translational modifications (PTM). Methods of monitoring PTM are an underdeveloped area in proteomics. Selecting a particular subset of protein for analysis substantially reduces protein complexity, making it advantageous for diagnostic purposes where blood is the starting material. Another important aspect of proteomics, yet not addressed, is that proteomics methods should focus on studying proteins in the context of the environment. The increasing use of chemical cross-linkers, introduced into living cells to fix protein-

protein, protein-DNA and other interactions, may ameliorate this problem partially. The challenge is to identify suitable methods of preserving relevant interactions. Another goal for studying proteins is development of more sophisticated methods to image proteins and other molecules in living cells and real-time.

Systems Biology

Advances in quantitative proteomics would clearly enable more in-depth analysis of cellular systems. Another research frontier is the analysis of single cells, and protein covariation across single cells which reflects biological processes such as protein complex formation, immune functions, as well as cell cycle and priming of cancer cells for drug resistance Biological systems are subject to a variety of perturbations (cell cycle, cellular differentiation, carcinogenesis, environment (biophysical), etc.).

Transcriptional and translational responses to these perturbations results in functional changes to the proteome implicated in response to the stimulus. Therefore, describing and quantifying proteome-wide changes in protein abundance is crucial towards understanding biological phenomenon more holistically, on the level of the entire system. In this way, proteomics can be seen as complementary to genomics, transcriptomics, epigenomics, metabolomics, and other -omics approaches in integrative analyses attempting to define biological phenotypes more comprehensively. As an example, *The Cancer Proteome Atlas* provides quantitative protein expression data for ~200 proteins in over 4,000 tumor samples with matched transcriptomic and genomic data from The Cancer Genome Atlas. Similar datasets in other cell types, tissue types, and species, particularly using deep shotgun mass spectrometry, will be an immensely important resource for research in fields like cancer biology, developmental and stem cell biology, medicine, and evolutionary biology.

Human Plasma Proteome

Characterizing the human plasma proteome has become a major goal in the proteomics arena, but it is also the most challenging proteomes of all human tissues. It contains immunoglobulin, cytokines, protein hormones, and secreted proteins indicative of infection on top of resident, hemostatic proteins. It also contains tissue leakage proteins due to the blood circulation through different tissues in the body. The blood thus contains information on the physiological state of all tissues and, combined with its accessibility, makes the blood proteome invaluable for medical purposes. It is thought that characterizing the proteome of blood plasma is a daunting challenge.

The depth of the plasma proteome encompasses a dynamic range of more than 10^{10} between the highest abundant protein (albumin) and the lowest (some cytokines) and is thought to be one of the main challenges for proteomics. Temporal and spatial dynamics further complicate the study of human plasma proteome. The turnover of some proteins is quite faster than others and the protein content of an artery may substantially vary from that of a vein. All these differences make even the simplest proteomic task of cataloging the proteome seem out of reach. To tackle this problem, priorities need to be established. Capturing the most meaningful subset of proteins among the entire proteome to generate a diagnostic tool is one such priority. Secondly, since cancer is associated with enhanced glycosylation of proteins, methods that focus on this part of proteins will also be useful. Again: multiparameter analysis best reveals a pathological state. As these technologies improve, the disease profiles should be continually related to respective gene expression changes. Due to the above-mentioned problems plasma proteomics remained challenging. However, technological advancements and continuous developments seem to result in a revival of plasma proteomics as it was shown recently by a technology called plasma proteome profiling. Due to such technologies researchers were able to investigate inflammation processes in mice, the heritability of plasma proteomes as well as to show the effect of such a common life style change like weight loss on the plasma proteome.

Metabolomics

Metabolomics is the scientific study of chemical processes involving metabolites, the small molecule substrates, intermediates, and products of cell metabolism. Specifically, metabolomics is the "systematic study of the unique chemical fingerprints that specific cellular processes leave behind", the study of their small-molecule metabolite profiles. The metabolome represents the complete set of metabolites in a biological cell, tissue, organ, or organism, which are the end products of cellular processes. Messenger RNA (mRNA), gene expression data and proteomic analyses reveal the set of gene products being produced in the cell, data that represents one aspect of cellular function. Conversely, metabolic profiling can give an instantaneous snapshot of the physiology of that cell, and thus, metabolomics provides a direct "functional readout of the physiological state" of an organism. There are indeed quantifiable correlations between the metabolome and the other cellular ensembles (genome, transcriptome, proteome, and lipidome), which can be used to predict metabolite abundances in biological samples from, for example mRNA abundances. One of the ultimate challenges of systems biology is to integrate metabolomics with all other -omics information to provide a better understanding of cellular biology.

History

- The concept that individuals might have a "metabolic profile" that could be reflected in the makeup of their biological fluids was introduced by Roger Williams in the late 1940s, who used paper chromatography to suggest characteristic metabolic patterns in urine and saliva were associated with diseases such as schizophrenia. However, it was only through technological advancements in the 1960s and 1970s that it became feasible to quantitatively (as opposed to qualitatively) measure metabolic profiles. The term "metabolic profile" was introduced by Horning, *et al.* in 1971 after they demonstrated that gas chromatography-mass spectrometry (GC-MS) could be used to measure compounds present in human urine and tissue extracts. The Horning group, along with that of Linus Pauling and Arthur B. Robinson led the development of GC-MS methods to monitor the metabolites present in urine through the 1970s.

- Concurrently, NMR spectroscopy, which was discovered in the 1940s, was also undergoing rapid advances. In 1974, Seeley et al. demonstrated the utility of using NMR to detect metabolites in unmodified biological samples. This first study on muscle highlighted the value of NMR in that it was determined that 90% of cellular ATP is complexed with magnesium. As sensitivity has improved with the evolution of higher magnetic field strengths and magic angle spinning, NMR continues to be a leading analytical tool to investigate metabolism. Recent efforts to utilize NMR for metabolomics have been largely driven by the laboratory of Jeremy K. Nicholson at Birkbeck College, University of London and later at Imperial College London. In 1984, Nicholson showed ^{1}H NMR spectroscopy could potentially be used to diagnose diabetes mellitus, and later pioneered the application of pattern recognition methods to NMR spectroscopic data.
- In 1994 and 1996, liquid chromatography mass spectrometry metabolomics experiments were performed by Gary Siuzdak while working with Richard Lerner (then president of the Scripps Research Institute) and Benjamin Cravatt, to analyze the cerebral spinal fluid from sleep deprived animals. One molecule of particular interest, oleamide, was observed and later shown to have sleep inducing properties. This work is one of the earliest such experiments combining liquid chromatography and mass spectrometry in metabolomics.
- In 2005, the first metabolomics tandem mass spectrometry database, METLIN, for characterizing human metabolites was developed in the Siuzdak laboratory at the Scripps Research Institute. METLIN has since grown and as of December, 2023, METLIN contains MS/MS experimental data on over 930,000 molecular standards and other chemical entities,

each compound having experimental tandem mass spectrometry data generated from molecular standards at multiple collision energies and in positive and negative ionization modes. METLIN is the largest repository of tandem mass spectrometry data of its kind. The dedicated academic journal Metabolomics first appeared in 2005, founded by its current editor-in-chief Roy Goodacre.

- In 2005, the Siuzdak lab was engaged in identifying metabolites associated with sepsis and in an effort to address the issue of statistically identifying the most relevant dysregulated metabolites across hundreds of LC/MS datasets, the first algorithm was developed to allow for the nonlinear alignment of mass spectrometry metabolomics data. Called XCMS, it has since (2012) been developed as an online tool and as of 2019 (with METLIN) has over 30,000 registered users.
- On 23 January 2007, the Human Metabolome Project, led by David S. Wishart, completed the first draft of the human metabolome, consisting of a database of approximately 2,500 metabolites, 1,200 drugs and 3,500 food components. Similar projects have been underway in several plant species, most notably *Medicago truncatula* and *Arabidopsis thaliana* for several years.
- As late as mid-2010, metabolomics was still considered an "emerging field". Further, it was noted that further progress in the field depended in large part, through addressing otherwise "irresolvable technical challenges", by technical evolution of mass spectrometry instrumentation.
- In 2015, real-time metabolome profiling was demonstrated for the first time.

Metabolome

The metabolome refers to the complete set of small-molecule (<1.5 kDa) metabolites (such as metabolic intermediates, hormones and other signaling molecules, and secondary metabolites) to be found within a biological sample, such as a single organism. The word was coined in analogy with transcriptomics and proteomics; like the transcriptome and the proteome, the metabolome is dynamic, changing from second to second. Although the metabolome can be defined readily enough, it is not currently possible to analyse the entire range of metabolites by a single analytical method.

In January 2007, scientists at the University of Alberta and the University of Calgary completed the first draft of the human metabolome. The Human Metabolome Database (HMDB) is perhaps the most extensive public metabolomic spectral database to date and is a freely available electronic database (www.hmdb.ca) containing detailed information about small molecule metabolites found in the human body. It is intended to be used for applications in metabolomics, clinical chemistry, biomarker discovery and general education. The database is designed to contain or link three kinds of data:

- Chemical data,
- Clinical data, and
- Molecular biology/biochemistry data.

The database contains 220,945 metabolite entries including both water-soluble and lipid soluble metabolites. Additionally, 8,610 protein sequences (enzymes and transporters) are linked to these metabolite entries. Each MetaboCard entry contains 130 data fields with 2/3 of the information being devoted to chemical/ clinical data and the other 1/3 devoted to enzymatic or biochemical data. The version 3.5 of the HMDB contains >16,000 endogenous metabolites, >1,500 drugs and >22,000 food constituents or food metabolites. This information, available at the Human Metabolome Database and based on analysis of information available in the current scientific literature, is far from complete. In contrast, much more is known about the metabolomes of other organisms. For example, over 50,000 metabolites have been characterized from the plant kingdom, and many thousands of metabolites have been identified and/or characterized from single plants.

Each type of cell and tissue has a unique metabolic 'fingerprint' that can elucidate organ or tissue-specific information. Bio-specimens used for metabolomics analysis include but not limit to plasma, serum, urine, saliva, feces, muscle, sweat, exhaled breath and gastrointestinal fluid. The ease of collection facilitates high temporal resolution, and because they are always at dynamic equilibrium with the body, they can describe the host as a whole. Genome can tell what could happen, transcriptome can tell what appears to be happening, proteome can tell what makes it happen and metabolome can tell what has happened and what is happening.

Metabolites

Metabolites are the substrates, intermediates and products of metabolism. Within the context of metabolomics, a metabolite is usually defined as any molecule less than 1.5 kDa in size. However, there are exceptions to this depending on the sample and detection method. For example, macromolecules such as lipoproteins and albumin are reliably detected in NMR-based metabolomics studies of blood plasma. In plant-based metabolomics, it is common to refer to "primary" and "secondary" metabolites. A primary metabolite is directly involved in the normal growth, development, and reproduction. A secondary metabolite is not directly involved in those processes, but usually has important ecological function. Examples include antibiotics and pigments. By contrast, in human-based metabolomics, it is more common to describe metabolites as being either endogenous (produced by the host organism) or exogenous. Metabolites of foreign substances such as drugs are termed xenometabolites. The metabolome forms a large network of metabolic reactions, where outputs from one enzymatic chemical reaction are inputs to other chemical reactions. Such systems have been described as hypercycles.

Metabonomics

Metabonomics is defined as "the quantitative measurement of the dynamic multiparametric metabolic response of living systems to pathophysiological stimuli or genetic modification". This approach was pioneered by Jeremy Nicholson at Murdoch University and has been used in toxicology, disease diagnosis and a number of other fields. Historically, the metabonomics approach was one of the first methods to apply the scope of systems biology to studies of metabolism.

There has been some disagreement over the exact differences between 'metabolomics' and 'metabonomics'. The difference between the two terms is not related to choice of analytical platform: although metabonomics is more associated with NMR spectroscopy and metabolomics with mass spectrometry-based techniques, this is simply because of usages amongst different groups that have popularized the different terms. While there is still no absolute agreement, there is a growing consensus that 'metabolomics' places a greater emphasis on metabolic profiling at a cellular or organ level and is primarily concerned with normal endogenous metabolism. 'Metabonomics' extends metabolic profiling to include information about perturbations of metabolism caused by environmental factors (including diet and toxins), disease processes, and the involvement of extragenomic influences, such as gut microflora. This is not a trivial difference metabolomic studies should, by definition, exclude metabolic contributions from extragenomic sources, because these are external to the system being studied. However, in practice, within the field of human disease research there is still a large degree of overlap in the way both terms are used, and they are often in effect synonymous.

Exometabolomics

Exometabolomics, or "metabolic footprinting", is the study of extracellular metabolites. It uses many techniques from other subfields of metabolomics, and has applications in biofuel development, bioprocessing, determining drugs' mechanism of action, and studying intercellular interactions.

Analytical Technologies

First, samples are collected from tissue, plasma, urine, saliva, cells, etc. Next, metabolites extracted often with the addition of internal standards and derivatization. During sample analysis, metabolites are quantified (liquid chromatography or gas chromatography coupled with MS and/or NMR spectroscopy). The raw output data can be used for metabolite feature extraction and further processed before statistical analysis (such as principal component analysis, PCA). Many bioinformatic tools and software are available to identify associations with disease states and outcomes, determine significant correlations, and characterize metabolic signatures with existing biological knowledge.

Separation Methods

Initially, analytes in a metabolomic sample comprise a highly complex mixture. This complex mixture can be simplified prior to detection by separating some analytes from others. Separation achieves various goals: analytes which cannot be resolved by the detector may be separated in this step; in MS analysis, ion suppression is reduced; the retention time of the analyte serves as information regarding its identity. This separation step is not mandatory and is often omitted in NMR and "shotgun" based approaches such as shotgun lipidomics.

Gas chromatography (GC), especially when interfaced with mass spectrometry (GC-MS), is a widely used separation technique for metabolomic analysis. GC offers very high chromatographic resolution, and can be used in conjunction with a flame ionization detector (GC/FID) or a mass spectrometer (GC-MS). The method is especially useful for identification and quantification of small and volatile molecules. However, a practical limitation of GC is the requirement of chemical derivatization for many biomolecules as only volatile chemicals can be analysed without derivatization. In cases where greater resolving power is required, two-dimensional chromatography (GCxGC) can be applied.

High performance liquid chromatography (HPLC) has emerged as the most common separation technique for metabolomic analysis. With the advent of electrospray ionization, HPLC was coupled to MS. In contrast with GC, HPLC has lower chromatographic resolution, but requires no derivatization for polar molecules, and separates molecules in the liquid phase. Additionally HPLC has the advantage that a much wider range of analytes can be measured with a higher sensitivity than GC methods.

Capillary electrophoresis (CE) has a higher theoretical separation efficiency than HPLC (although requiring much more time per separation), and is suitable for use with a wider range of metabolite classes than is GC. As for all electrophoretic techniques, it is most appropriate for charged analytes.

Detection Methods

Mass spectrometry (MS) is used to identify and quantify metabolites after optional separation by GC, HPLC, or CE. GC-MS was the first hyphenated technique to be developed. Identification leverages the distinct patterns in which analytes fragment. These patterns can be thought of as a mass spectral fingerprint. Libraries exist that allow identification of a metabolite according to this fragmentation pattern . MS is both sensitive and can be very specific. There are also a number of techniques which use MS as a stand-alone technology: the sample is infused directly into the mass spectrometer with no prior separation, and the MS provides sufficient selectivity to both separate and to detect metabolites.

For analysis by mass spectrometry, the analytes must be imparted with a charge and transferred to the gas phase. Electron ionization (EI) is the most common ionization technique applied to GC separations as it is amenable to low pressures. EI also produces fragmentation of the analyte, both providing structural information while increasing the complexity of the data and possibly obscuring the molecular ion. Atmospheric-pressure chemical ionization (APCI) is an atmospheric pressure technique that can be applied to all the above separation techniques. APCI is a gas phase ionization method, which provides slightly more aggressive ionization than ESI which is suitable for less polar compounds. Electrospray ionization (ESI) is the most common ionization technique applied in LC/MS. This soft ionization is most successful for polar molecules with ionizable functional groups. Another commonly used soft ionization technique is secondary electrospray ionization (SESI).

In the 2000s, surface-based mass analysis has seen a resurgence, with new MS technologies focused on increasing sensitivity, minimizing background, and reducing sample preparation. The ability to analyze metabolites directly from biofluids and tissues continues to challenge current MS technology, largely because of the limits imposed by the complexity of these samples, which contain thousands to tens of thousands of metabolites. Among the technologies being developed to address this challenge is Nanostructure-Initiator MS (NIMS), a desorption/ionization approach that does not require the application of matrix and thereby facilitates small-molecule (i.e., metabolite) identification. MALDI is also used; however, the application of a MALDI matrix can add significant background at < 1000 Da that complicates analysis of the low-mass range (i.e., metabolites). In addition, the size of the resulting matrix crystals limits the spatial resolution that can be achieved in tissue imaging. Because of these limitations, several other matrix-free desorption/ionization approaches have been applied to the analysis of biofluids and tissues.

Secondary ion mass spectrometry (SIMS) was one of the first matrix-free desorption/ionization approaches used to analyze metabolites from biological samples. SIMS uses a high-energy primary ion beam to desorb and generate secondary ions from a surface. The primary advantage of SIMS is its high spatial resolution (as small as 50 nm), a powerful characteristic for tissue imaging with MS. However, SIMS has yet to be readily applied to the analysis of biofluids and tissues because of its limited sensitivity at >500 Da and analyte fragmentation generated by the high-energy primary ion beam. Desorption electrospray ionization (DESI) is a matrix-free technique for analyzing biological samples that uses a charged solvent spray to desorb ions from a surface. Advantages of DESI are that no special surface is required and the analysis is performed at ambient pressure with full access to the sample during acquisition. A limitation of DESI is spatial resolution because "focusing" the charged solvent spray is difficult. However, a recent development termed laser ablation ESI (LAESI) is a promising approach to circumvent this limitation. Most recently, ion trap techniques such as orbitrap mass spectrometry are also applied to metabolomics research.

Nuclear magnetic resonance (NMR) spectroscopy is the only detection technique which does not rely on separation of the analytes, and the sample can thus be recovered for further analyses. All kinds of small molecule metabolites can be measured simultaneously - in this sense, NMR is close to being a universal detector. The main advantages of NMR are high analytical reproducibility and simplicity of sample preparation. Practically, however, it is relatively insensitive compared to mass spectrometry-based techniques.

Statistical Methods

The data generated in metabolomics usually consist of measurements performed on subjects under various conditions. These measurements may be digitized spectra, or a list of metabolite features. In its simplest form, this generates a matrix with rows corresponding to subjects and columns corresponding with metabolite features (or vice versa). Several statistical programs are currently available for analysis of both NMR and mass spectrometry data. For mass spectrometry data, software is available that identifies molecules that vary in subject groups on the basis of mass-over-charge value and sometimes retention time depending on the experimental design.

Once metabolite data matrix is determined, unsupervised data reduction techniques (e.g. PCA) can be used to elucidate patterns and connections. In many studies, including those evaluating drug-toxicity and some disease models, the metabolites of interest are not known *a priori*. This makes unsupervised methods, those with no prior assumptions of class membership, a popular first choice. The most common of these methods includes principal component analysis (PCA) which can efficiently reduce the dimensions of a dataset to a few which explain the greatest variation. When analyzed in the lower-dimensional PCA space, clustering of samples with similar metabolic fingerprints can be detected. PCA algorithms aim to replace all correlated variables with a much smaller number of uncorrelated variables (referred to as principal components (PCs)) and retain most of the information in the original dataset. This clustering can elucidate patterns and assist in the determination of disease biomarkers – metabolites that correlate most with class membership. Linear models are commonly used for metabolomics data, but are affected by multicollinearity. On the other hand, multivariate statistics are thriving methods for high-dimensional correlated metabolomics data, of which the most popular one is Projection to Latent Structures (PLS) regression and its classification version PLS-DA. Other data mining methods, such as random forest, support-vector machines, etc. are received increasing attention for untargeted metabolomics data analysis. In the case of univariate methods, variables are analyzed one by one using classical statistics tools (such as Student's t-test, ANOVA or mixed models) and only these with sufficient small p-values are considered relevant. However, correction strategies should be used to reduce false discoveries when multiple comparisons are conducted since there is no standard method for measuring the total amount of metabolites directly in untargeted metabolomics. For multivariate analysis, models should always be validated to ensure that the results can be generalized.

Machine Learning and Data Mining

Machine learning is a powerful tool that can be used in metabolomics analysis. Recently, scientists have developed retention time prediction software. These tools allow researchers to apply artificial intelligence to the retention time prediction of small molecules in complex mixture, such as human plasma, plant extracts, foods, or microbial cultures. Retention time prediction increases the identification rate in liquid chromatography and can lead to an improved biological interpretation of metabolomics data.

Key Applications

Toxicity assessment/toxicology by metabolic profiling (especially of urine or blood plasma samples) detects the physiological changes caused by toxic insult of a chemical (or mixture of chemicals). In many cases, the observed changes can be related to specific syndromes, e.g. a specific lesion in liver or kidney. This is of particular relevance to pharmaceutical companies wanting to test the toxicity of potential drug candidates: if a compound can be eliminated before it reaches clinical trials on the grounds of adverse toxicity, it saves the enormous expense of the trials. For functional genomics, metabolomics can be an excellent tool for determining the phenotype caused by a genetic manipulation, such as gene deletion or insertion. Sometimes this can be a sufficient goal in itself—for instance, to detect any phenotypic changes in a genetically modified plant intended for human or animal consumption. More exciting is the prospect of predicting the function of unknown genes by comparison with the metabolic perturbations caused by deletion/insertion of known genes. Such advances are most likely to come from model organisms such as *Saccharomyces cerevisiae* and *Arabidopsis thaliana.* The Cravatt laboratory at the Scripps Research Institute has recently applied this technology to mammalian systems, identifying the *N*-acyltaurines as previously uncharacterized endogenous substrates for the enzyme fatty acid amide hydrolase (FAAH) and the monoalkylglycerol ethers (MAGEs) as endogenous substrates for the uncharacterized hydrolase KIAA1363.

Metabologenomics is a novel approach to integrate metabolomics and genomics data by correlating microbial-exported metabolites with predicted biosynthetic genes. This bioinformatics-based pairing method enables natural product discovery at a larger-scale by refining non-targeted metabolomic analyses to identify small molecules with related biosynthesis and to focus on those that may not have previously well known structures.

Fluxomics is a further development of metabolomics. The disadvantage of metabolomics is that it only provides the user with abundances or concentrations of metabolites, while fluxomics determines the reaction rates of metabolic reactions and can trace metabolites in a biological system over time.

Nutrigenomics is a generalised term which links genomics, transcriptomics, proteomics and metabolomics to human nutrition. In general, in a given body fluid, a metabolome is influenced by endogenous factors such as age, sex, body composition and genetics as well as underlying pathologies. The large bowel microflora are also a very significant potential confounder of metabolic profiles and could be classified as either an endogenous or exogenous factor. The main exogenous factors are diet and drugs. Diet can then be broken down to nutrients and non-nutrients. Metabolomics is one means to determine a biological endpoint, or metabolic fingerprint, which reflects the balance of all these forces on an individual's metabolism. Thanks to recent cost reductions, metabolomics has now become accessible for companion animals, such as pregnant dogs.

Plant metabolomics is designed to study the overall changes in metabolites of plant samples and then conduct deep data mining and chemometric analysis. Specialized metabolites are considered components of plant defense systems biosynthesized in response to biotic and abiotic stresses. Metabolomics approaches have recently been used to assess the natural variance in metabolite content between individual plants, an approach with great potential for the improvement of the compositional quality of crops.

14.3 TOOLS AND TECHNOLOGIES IN FUNCTIONAL GENOMICS

14.3.1 Bioinformatics and Computational Tools

The integration of bioinformatics tools is essential for analyzing the vast amounts of data generated by functional genomics studies. Software platforms for genomic data analysis, such as Galaxy and Bioconductor, allow researchers to process and interpret sequencing data efficiently. These tools enable the identification of differentially expressed genes, functional annotations, and pathway analyses, facilitating the understanding of complex biological systems.

Additionally, machine learning algorithms are increasingly being applied in functional genomics to predict gene functions, interactions, and regulatory networks. These computational approaches enhance the ability to derive meaningful insights from genomic data, leading to more informed decisions in strain optimization and product development.

14.3.2 CRISPR and Genome Editing Technologies

The development of CRISPR-Cas9 technology has revolutionized functional genomics by providing a powerful tool for targeted gene editing. This genome editing technique allows researchers to modify specific genes, enabling the investigation of their functions and the creation of engineered microbial strains with desired traits. In the fermentation industry, CRISPR can be used to enhance the production of biofuels, enzymes, and pharmaceuticals by knocking out undesirable pathways or introducing beneficial genes. Moreover, CRISPR-based approaches are being employed to explore the genetic basis of traits such as stress tolerance, substrate utilization, and metabolite production. By elucidating the roles of specific genes, researchers can develop more robust microbial strains that perform efficiently under industrial fermentation conditions.

14.3.3 Systems Biology Approaches

Systems biology integrates functional genomics data with other omics technologies to create comprehensive models of biological systems. By analyzing the interactions between genes, proteins, and metabolites, researchers can gain insights into the regulatory networks that govern fermentation processes. Systems biology approaches can be employed to optimize metabolic pathways, predict the effects of genetic modifications, and identify potential bottlenecks in product formation.

In the pharmaceutical industry, systems biology can facilitate drug discovery by elucidating the molecular mechanisms underlying disease processes. By integrating functional genomics data with clinical and pharmacological information, researchers can identify novel drug targets and develop more effective therapeutic strategies.

14.4 FUTURE PROSPECTS IN FUNCTIONAL GENOMICS

14.4.1 Personalized Medicine and Drug Development

The future of functional genomics holds significant promise for the pharmaceutical industry, particularly in the realm of personalized medicine. By understanding the genetic basis of individual responses to drugs, researchers can develop tailored therapeutic approaches that maximize efficacy and minimize adverse effects. Functional genomics will play a pivotal role in identifying biomarkers for patient stratification, enabling more precise drug development and clinical decision-making.

Moreover, the integration of functional genomics with pharmacogenomics—the study of how genes influence drug response—will pave the way for the development of targeted therapies. By elucidating the molecular mechanisms underlying drug action and resistance, researchers can design drugs that are more effective for specific patient populations.

14.4.2 Sustainable Fermentation Processes

As the demand for sustainable production methods grows, functional genomics will be instrumental in optimizing fermentation processes for renewable resource utilization. By engineering microbial strains with enhanced metabolic capabilities, researchers can improve the efficiency of converting biomass into biofuels, chemicals, and pharmaceuticals. The integration of functional genomics with synthetic biology will enable the design of novel metabolic pathways and the production of high-value bioproducts from lignocellulosic biomass.

Furthermore, functional genomics can aid in developing strategies for reducing waste and by-products in fermentation processes. By understanding the regulatory networks that control metabolic flux, researchers can optimize fermentation conditions to enhance desired product formation while minimizing undesirable by-products.

14.4.3 Global Collaborations and Data Sharing

The future of functional genomics will increasingly rely on global collaborations and data-sharing initiatives. By pooling resources and expertise, researchers can accelerate the pace of discovery and innovation in both fermentation and pharmaceutical applications. Open-access databases and collaborative platforms will facilitate the exchange of genomic data, enabling researchers to build on each other's findings and drive advancements in the field.

14.5 CONCLUSION

Functional genomics has emerged as a transformative force in the fermentation and pharmaceutical industries, providing valuable insights into gene function and metabolic processes. Current trends, including high-throughput sequencing, transcriptomics, and systems biology approaches, are revolutionizing our understanding of microbial systems and enabling the development of engineered strains with enhanced production capabilities. As we look to the future, the integration of functional genomics with personalized medicine, sustainable fermentation processes, and global collaborations will shape the next generation of innovations in these industries. By harnessing the power of functional genomics, researchers can unlock new avenues for product development, improve efficiency, and contribute to a more sustainable and health-conscious future.

CHAPTER 15

BIOREACTORS: DESIGN, OPERATION, AND APPLICATIONS

Bioreactors are vital components in biotechnology and industrial microbiology, serving as controlled environments where biological reactions take place. They enable the cultivation of microorganisms, plant cells, or animal cells to produce various products, including enzymes, pharmaceuticals, and biofuels. The design and operation of bioreactors are critical factors influencing the efficiency and yield of these processes.

15.1 UNDERSTANDING BIOREACTORS

A bioreactor is a vessel or container designed to provide a suitable environment for biological processes. These processes often involve the growth of microorganisms or the production of biological compounds. The primary goal of a bioreactor is to optimize the conditions under which these biological systems operate, ensuring maximum productivity while maintaining the health and viability of the cells involved. Bioreactors can be classified into various types based on their operational modes, scale, and biological systems utilized. Common categories include batch, fed-batch, and continuous bioreactors. Each type has distinct advantages and challenges, influencing their selection for specific applications.

15.2 DESIGN PRINCIPLES OF BIOREACTORS

15.2.1 Vessel Design

The design of the bioreactor vessel is crucial for effective mixing, mass transfer, and heat dissipation. The shape, size, and materials used in the construction of the bioreactor can significantly impact its performance. Typically, bioreactors are cylindrical with a flat or conical bottom, which facilitates uniform mixing and ensures effective nutrient distribution. The material of construction is also important, as it must be compatible with the biological agents being used and resistant to corrosion and degradation. Common materials include stainless steel and glass, which provide durability and ease of cleaning.

15.2.2 Mixing and Mass Transfer

Effective mixing is essential in a bioreactor to ensure that nutrients, gases, and cells are evenly distributed throughout the culture medium. Various mixing mechanisms are employed, such as mechanical agitation using impellers, sparging with gases, or using static mixers. The choice of mixing method depends on the type of bioreactor, the sensitivity of the organisms being cultured, and the desired level of shear stress.

Mass transfer is another critical aspect, particularly in aerobic fermentations where oxygen transfer is vital. The design must allow for optimal gas-liquid interactions to maintain sufficient dissolved oxygen levels. This may involve using specialized aeration systems or increasing the surface area for gas exchange.

15.2.3 Temperature and pH Control

Maintaining optimal temperature and pH conditions is crucial for the growth and productivity of microorganisms. Bioreactors are equipped with temperature control systems, such as heating jackets or cooling coils, to regulate the temperature within the vessel. Additionally, pH control systems, often involving the addition of acids or bases, are implemented to maintain the desired pH range. Sensors and automated control systems are typically used to monitor and adjust these parameters in real time.

15.3 OPERATION OF BIOREACTORS

15.3.1 Sterilization

Before inoculation, bioreactors must be sterilized to eliminate any unwanted microorganisms that could compete with the desired cells. This is usually achieved through steam sterilization (autoclaving) or chemical sterilization methods. Ensuring aseptic conditions is critical for the success of the fermentation process and the quality of the final product.

15.3.2 Inoculation and Cultivation

After sterilization, the bioreactor is inoculated with the desired microorganism or cell line. The cultivation phase involves monitoring and controlling various parameters, including nutrient supply, oxygen levels, temperature, and pH, to create optimal conditions for growth. Depending on the operational mode, nutrient feeding strategies may vary. In batch systems, all nutrients are provided initially, while fed-batch systems involve the continuous addition of nutrients throughout the cultivation process.

15.3.3 Harvesting

The harvesting phase involves extracting the desired products from the bioreactor after the fermentation process is complete. This can include the separation of cells, purification of metabolites, or recovery of proteins. Various techniques, such as centrifugation, filtration, and chromatography, are employed to isolate and purify the target compounds effectively.

15.4 APPLICATIONS OF BIOREACTORS

Bioreactors have a wide range of applications across various industries, particularly in pharmaceuticals, agriculture, food and beverage, and environmental biotechnology.

15.4.1 Pharmaceutical Industry

In the pharmaceutical sector, bioreactors play a crucial role in the production of biopharmaceuticals, including monoclonal antibodies, vaccines, and recombinant proteins. Large-scale bioreactors are utilized to culture mammalian or microbial cells that produce these therapeutic agents. The ability to control environmental conditions ensures consistent quality and yield, which are critical for regulatory compliance in pharmaceutical manufacturing.

15.4.2 Food and Beverage Industry

Bioreactors are extensively used in the food and beverage industry for the production of fermented products such as yogurt, cheese, beer, and wine. In this context, bioreactors facilitate the controlled fermentation process, ensuring the desired flavor, aroma, and texture of the final product. The precise control of temperature, pH, and oxygen levels contributes to the quality and safety of fermented foods.

15.4.3 Biofuels Production

The production of biofuels, such as ethanol and biodiesel, increasingly relies on bioreactors for the fermentation of biomass. Microorganisms convert sugars from plant materials into alcohol or other fuel sources, providing a renewable alternative to fossil fuels. Continuous bioreactors are particularly suited for biofuel production, allowing for sustained operation and higher yields.

15.4.4 Environmental Biotechnology

Bioreactors also find applications in environmental biotechnology, particularly in wastewater treatment and bioremediation. In these contexts, bioreactors can facilitate the degradation of pollutants by harnessing the metabolic capabilities of microorganisms. Engineered bioreactors, such as sequencing batch reactors (SBR) or membrane bioreactors (MBR), are employed to treat wastewater efficiently, reducing environmental impact and promoting resource recovery.

15.5 CHALLENGES IN BIOREACTOR DESIGN AND OPERATION

Despite their numerous advantages, bioreactors face several challenges that can impact their performance and efficiency. One of the primary challenges is the scale-up of processes from laboratory to industrial scales. Maintaining consistent conditions and performance during scale-up requires careful consideration of mixing, mass transfer, and heat dissipation.

Another challenge is the potential for contamination, which can compromise product quality and yield. Implementing strict aseptic practices and monitoring systems is essential to minimize contamination risks.

Additionally, optimizing bioreactor design and operation for specific applications may involve significant experimentation and time. Researchers and engineers must balance multiple parameters, including economic feasibility, environmental sustainability, and product quality.

15.6 FUTURE PROSPECTS IN BIOREACTOR TECHNOLOGY

The future of bioreactor technology holds significant promise as advancements in materials science, automation, and data analytics continue to evolve. Innovations such as smart bioreactors equipped with sensors and IoT connectivity will enable real-time monitoring and control of fermentation processes. This level of automation can enhance operational efficiency and reduce human error, ultimately leading to improved product quality.

Moreover, the integration of machine learning and artificial intelligence in bioreactor design and operation is anticipated to revolutionize the industry. Predictive models can optimize operational parameters and facilitate rapid adjustments based on real-time data, enabling more efficient and robust fermentation processes.

As sustainability becomes a central focus across industries, bioreactors that utilize renewable resources and minimize waste will be increasingly prioritized. The development of bioreactors that support circular economy principles will enhance resource recovery and reduce environmental impact.

15.7 CONCLUSION

Bioreactors are integral to the success of various biotechnological processes, enabling the controlled cultivation of microorganisms and the production of valuable products across numerous industries. Their design and operation are critical factors that influence productivity, product quality, and process efficiency. As advancements in technology and a growing focus on sustainability drive innovation in bioreactor design and operation, the future prospects for bioreactors appear bright. By continuing to optimize bioreactor systems, the biotechnology and pharmaceutical industries can unlock new opportunities for sustainable production and address the challenges of tomorrow.

CHAPTER 16

BIOSENSORS IN BIOPROCESS MONITORING AND CONTROL: CURRENT TRENDS AND FUTURE PROSPECTS

Biosensors have emerged as crucial tools in bioprocess monitoring and control, offering real-time analysis of biological systems. These devices combine biological components with a physicochemical detector to quantify various parameters, such as metabolites, enzymes, and cellular activity, facilitating more efficient and reliable bioprocesses.

16.1 UNDERSTANDING BIOSENSORS

Biosensors are analytical devices that convert a biological response into an electrical signal. They typically consist of three main components: a biorecognition element, a transducer, and a signal processor. The biorecognition element can be enzymes, antibodies, or whole cells, which interact specifically with the target analyte. The transducer converts this interaction into a measurable signal, often in the form of current, voltage, or optical changes. Finally, the signal processor interprets the signal, providing quantitative or qualitative data about the analyte's concentration. The advantages of biosensors include their high specificity, sensitivity, and ability to provide real-time data. These features make them invaluable for monitoring bioprocesses in various applications, including pharmaceuticals, food production, and environmental monitoring.

16.2 CURRENT TRENDS IN BIOSENSOR TECHNOLOGY

16.2.1 Miniaturization and Portability

One of the most significant trends in biosensor development is the miniaturization of devices, enabling portability and ease of use. Advances in microfabrication techniques have allowed researchers to create compact biosensors that can be used in field applications or integrated into portable diagnostic devices. This trend is particularly relevant in bioprocess monitoring, where real-time data collection is essential for maintaining optimal conditions. Portable biosensors can be used in various settings, from laboratory-scale experiments to large industrial bioreactors.

Their ability to provide immediate feedback allows for rapid decision-making and adjustments in bioprocess parameters, enhancing overall efficiency.

16.2.2 Integration with Smart Technologies

The integration of biosensors with smart technologies, such as the Internet of Things (IoT) and artificial intelligence (AI), is transforming bioprocess monitoring. IoT-enabled biosensors can connect to cloud-based platforms, allowing for remote monitoring and data analysis.

This connectivity enables real-time data sharing and collaboration among researchers, enhancing the overall management of bioprocesses. Moreover, AI algorithms can analyze the data collected by biosensors, identifying patterns and making predictions about bioprocess performance. This predictive capability can facilitate proactive adjustments to fermentation conditions, leading to improved yield and reduced downtime.

16.2.3 Multi-Analyte Detection

Another trend in biosensor technology is the development of multi-analyte sensors capable of simultaneously measuring multiple parameters. Traditional biosensors often focus on a single analyte, limiting their application in complex bioprocesses where various factors must be monitored. Multi-analyte sensors can provide a comprehensive overview of the bioprocess environment, including key metabolites, pH, temperature, and dissolved oxygen levels.

These devices enhance bioprocess control by allowing for the correlation of different parameters, leading to a better understanding of the interactions that influence microbial growth and product formation. This holistic approach is crucial for optimizing bioprocesses in real time.

16.3 APPLICATIONS OF BIOSENSORS IN BIOPROCESS MONITORING AND CONTROL

16.3.1 Fermentation Monitoring

Biosensors play a vital role in monitoring fermentation processes, where precise control of parameters is crucial for achieving high product yields. For instance, glucose biosensors can continuously monitor sugar levels in fermentation broths, enabling operators to adjust nutrient feed rates accordingly. Similarly, sensors that detect ethanol or organic acids can provide real-time data on product concentrations, allowing for timely adjustments to fermentation conditions.

By integrating biosensors into fermentation systems, operators can achieve better control over metabolic pathways, minimizing by-product formation and maximizing desired product yield.

16.3.2 Cell Culture Monitoring

In the pharmaceutical industry, biosensors are increasingly used to monitor mammalian cell cultures for the production of biopharmaceuticals. These sensors can detect critical parameters such as glucose, lactate, and ammonia levels, providing insights into cell metabolism and health. Monitoring these parameters is essential for maintaining optimal growth conditions and preventing cell death. Real-time data from biosensors can help optimize feeding strategies, leading to improved cell viability and productivity. Additionally, biosensors can monitor the expression of specific proteins, ensuring the quality of the final biopharmaceutical product.

16.3.3 Environmental Monitoring

Biosensors are also applied in environmental monitoring, where they can detect pollutants, pathogens, and other harmful substances in water and soil. For example, biosensors that utilize microorganisms as biorecognition elements can identify specific contaminants, providing valuable information for environmental remediation efforts. In bioprocesses that involve wastewater treatment, biosensors can monitor the concentration of organic matter and other parameters, ensuring that treatment processes are functioning effectively. This application not only improves efficiency but also contributes to sustainable environmental practices.

16.4 FUTURE PROSPECTS IN BIOSENSOR TECHNOLOGY

16.4.1 Advances in Nanotechnology

The integration of nanotechnology into biosensor design holds great promise for enhancing sensitivity and specificity. Nanomaterials, such as nanoparticles and carbon nanotubes, can be used to improve the performance of transducers, leading to more accurate and reliable measurements. These materials can enhance the surface area and reactivity of biosensors, enabling the detection of lower concentrations of analytes. Future biosensors that leverage nanotechnology may be capable of detecting multiple analytes simultaneously with greater precision, further improving their utility in bioprocess monitoring.

16.4.2 Personalized Bioprocess Control

As biosensor technology advances, the potential for personalized bioprocess control becomes more feasible. By integrating biosensors with machine learning algorithms, bioprocesses can be tailored to the specific needs of different microbial strains or cell lines. This personalized approach could lead to higher product yields and improved process stability. Additionally, personalized monitoring could enable the development of adaptive bioprocesses that respond dynamically to changes in environmental conditions or biological activity, resulting in more robust production systems.

16.4.3 Regulatory Acceptance and Standardization

For biosensors to be widely adopted in bioprocess monitoring and control, regulatory acceptance and standardization are essential. As biosensor technology matures, regulatory bodies are likely to develop guidelines and frameworks for the validation and certification of biosensors used in industrial applications. This acceptance will enhance confidence in the reliability of biosensors and facilitate their integration into established bioprocesses. Standardization of biosensor performance metrics and data reporting will also be crucial for ensuring compatibility and interoperability across different systems and applications.

16.5 CONCLUSION

Biosensors have become indispensable tools in the monitoring and control of bioprocesses, providing real-time data that enhances efficiency and product quality. Current trends, such as miniaturization, integration with smart technologies, and multi-analyte detection, are transforming the landscape of biosensor applications in industries ranging from pharmaceuticals to environmental monitoring. As technology continues to advance, the future of biosensors looks promising, with potential developments in nanotechnology, personalized bioprocess control, and regulatory acceptance. By harnessing the capabilities of biosensors, industries can optimize their processes, reduce costs, and contribute to a more sustainable future.

CHAPTER 17

CELL IMMOBILIZATION AND ITS APPLICATIONS IN BIOTECHNOLOGY: CURRENT TRENDS AND FUTURE PROSPECTS

Cell immobilization is a crucial technique in biotechnology that involves the confinement of cells within a defined space, enabling their use in various bioprocesses. This method enhances the stability, efficiency, and longevity of biocatalysts, making it an essential practice in the production of pharmaceuticals, biofuels, and other bioproducts.

17.1 UNDERSTANDING CELL IMMOBILIZATION

Cell immobilization refers to the process of restricting the movement of cells within a specific environment while maintaining their metabolic activity. This can be achieved through various methods, including physical entrapment, covalent bonding, and adsorption onto solid supports. The primary goal of cell immobilization is to enhance the efficiency of biocatalysts by allowing for continuous operation, ease of separation from the reaction mixture, and improved stability under various conditions. Immobilized cells can be used in numerous biotechnological applications, providing significant advantages over free cells, such as higher concentration of active biomass, reduced product inhibition, and the potential for repeated use of the same culture. These advantages make cell immobilization a vital technique in the field of biotechnology.

17.2 CURRENT TRENDS IN CELL IMMOBILIZATION

17.2.1 Development of Advanced Immobilization Techniques

Recent advancements in immobilization techniques have led to the development of novel methods that enhance cell stability and activity. Traditional methods, such as adsorption and entrapment, are still widely used; however, new techniques like covalent bonding and the use of advanced materials are gaining popularity. For instance, the use of nanomaterials, hydrogels, and composite matrices provides improved support for immobilized cells, enhancing their performance in various bioprocesses. Furthermore, researchers are exploring methods that enable the selective immobilization of specific cell types, allowing for tailored biocatalytic processes. This specificity can significantly improve the efficiency of reactions, making them more cost-effective.

17.2.2 Integration with Microbial Fuel Cells (MFCs)

Microbial fuel cells (MFCs) are an emerging technology that harnesses the metabolic activity of immobilized microorganisms to convert organic substrates into electricity. The integration of cell immobilization techniques into MFCs has shown promising results, leading to improved power generation efficiency and stability. Immobilized cells provide a stable source of electrons, enhancing the overall performance of the fuel cell. Recent studies have focused on optimizing the design of MFCs by selecting suitable immobilization techniques and materials. These advancements are expected to pave the way for more sustainable energy solutions, utilizing waste products as substrates for electricity generation.

17.2.3 Enhanced Application of Enzymes through Immobilization

Enzyme immobilization is a critical area of focus within cell immobilization research. The ability to enhance enzyme stability, reusability, and activity through immobilization has made it an essential technique in various industrial applications. Recent trends include the development of enzyme carriers, such as magnetic nanoparticles and polymeric supports, which allow for easy separation and recovery of enzymes after use. Moreover, advancements in enzyme engineering, coupled with immobilization techniques, are leading to the creation of tailored enzymes with improved catalytic properties. This trend is particularly significant in the food, pharmaceutical, and biofuel industries, where immobilized enzymes can significantly enhance production efficiency.

17.3 APPLICATIONS OF CELL IMMOBILIZATION IN BIOTECHNOLOGY

17.3.1 Bioremediation

Cell immobilization has found extensive applications in bioremediation, the process of using microorganisms to degrade environmental pollutants. Immobilized cells can be used to remove contaminants from soil and water, providing a more efficient and sustainable approach to environmental cleanup. For instance, immobilized bacteria have been employed to degrade heavy metals, pesticides, and petroleum hydrocarbons.

The use of immobilized cells enhances their stability and activity, allowing them to effectively remove pollutants from contaminated sites over extended periods. This application is crucial for addressing environmental challenges and promoting sustainable practices.

17.3.2 Production of Biofuels

The production of biofuels, such as ethanol and biodiesel, is another significant area where cell immobilization plays a vital role. Immobilized microorganisms, such as yeast and bacteria, can be used to convert biomass into biofuels, offering a more efficient and cost-effective alternative to traditional fermentation processes. By immobilizing the microorganisms, it becomes easier to separate them from the fermentation broth, allowing for continuous operation and increased productivity. Furthermore, immobilized cells exhibit enhanced tolerance to inhibitors commonly found in biomass hydrolysates, making them more effective in biofuel production processes.

17.3.3 Pharmaceutical Manufacturing

In the pharmaceutical industry, cell immobilization is employed to enhance the production of various biopharmaceuticals, including enzymes, antibodies, and vaccines. Immobilized cells can be utilized in bioreactors for the continuous production of these products, improving overall efficiency and reducing production costs. The ability to recycle and reuse immobilized cells significantly contributes to the sustainability of pharmaceutical manufacturing processes. Moreover, advancements in immobilization techniques have enabled the development of engineered cell lines that can produce high yields of specific biopharmaceuticals, meeting the increasing demand for these products in healthcare.

17.4 FUTURE PROSPECTS IN CELL IMMOBILIZATION

17.4.1 Advances in Synthetic Biology

The field of synthetic biology is poised to revolutionize cell immobilization techniques by enabling the design and construction of custom microorganisms with desired characteristics. This approach allows for the development of tailored biocatalysts that can be effectively immobilized, enhancing their performance in various biotechnological applications. Furthermore, synthetic biology can facilitate the integration of biosensors into immobilized cell systems, enabling real-time monitoring of metabolic activity and product formation. This advancement could lead to more efficient and responsive bioprocesses, significantly improving production yields.

17.4.2 Development of Smart Immobilization Systems

Future trends in cell immobilization may involve the creation of smart immobilization systems capable of adapting to changing environmental conditions. These systems could utilize responsive materials that change properties based on external stimuli, allowing for controlled release or enhanced activity of immobilized cells. Such innovations would enable more precise control over bioprocesses, optimizing performance while minimizing resource consumption. The development of smart immobilization systems could have far-reaching implications in industries such as food production, pharmaceuticals, and environmental biotechnology.

17.4.3 Enhanced Sustainability through Circular Economy Principles

As the global focus on sustainability intensifies, the application of cell immobilization techniques can contribute to the principles of a circular economy. By enabling the efficient reuse of biological materials and reducing waste, cell immobilization aligns with the growing demand for sustainable biotechnological practices. Future developments may include the integration of immobilized cells in closed-loop systems, where by-products from one process can be utilized as substrates for another. This interconnected approach has the potential to enhance overall efficiency and reduce environmental impact, creating a more sustainable framework for biotechnological applications.

17.5 CONCLUSION

Cell immobilization is a vital technique in biotechnology that offers numerous advantages for enhancing the efficiency and stability of biocatalysts. Current trends highlight advancements in immobilization techniques, the integration of biosensors, and applications in areas such as bioremediation, biofuel production, and pharmaceutical manufacturing. As the field continues to evolve, future prospects in cell immobilization hold great promise, particularly with the integration of synthetic biology, smart systems, and sustainability principles. By harnessing the power of cell immobilization, the biotechnology industry can develop innovative solutions to address global challenges and promote sustainable practices for the future.

CHAPTER 18

RECENT ADVANCES AND IMPACTS OF MICROTITER PLATE-BASED FERMENTATIONS IN SYNTHETIC BIOLOGY AND BIOPROCESS DEVELOPMENT

Microtiter plates (MTPs) have revolutionized the landscape of biological research, providing a high-throughput platform for various applications, including synthetic biology and bioprocess development. These multi-well plates, typically containing 96, 384, or even 1536 wells, allow for the parallel screening of numerous conditions, organisms, and genetic constructs.

18.1 UNDERSTANDING MICROTITER PLATE-BASED FERMENTATION

Microtiter plate-based fermentation refers to the use of microtiter plates to conduct fermentation processes at a small scale. These plates allow researchers to culture microorganisms, perform metabolic studies, and screen for optimal conditions in a highly efficient manner. The advantages of MTPs include reduced reagent usage, minimal culture volume, and the ability to monitor multiple parameters simultaneously. These features make them particularly valuable in synthetic biology, where rapid iteration and optimization are critical. Microtiter plates are equipped with various configurations, including different well sizes, shapes, and volumes, allowing researchers to tailor their experiments to specific needs. The use of automation and advanced analytical techniques further enhances the utility of MTPs in fermentation studies, facilitating real-time monitoring and data collection.

18.2 RECENT ADVANCES IN MICROTITER PLATE-BASED FERMENTATIONS

18.2.1 Automation and High-Throughput Screening

One of the most significant advances in microtiter plate-based fermentations is the automation of processes, which enables high-throughput screening of microbial strains and conditions. Automated liquid handling systems can quickly dispense culture media, inoculate wells, and add reagents, drastically reducing manual labor and the potential for human error. High-throughput screening allows researchers to evaluate numerous variables simultaneously, including nutrient concentrations, pH levels, and temperature conditions. This capability is crucial for synthetic biology applications, where optimizing metabolic pathways and gene expression often requires extensive experimentation. Recent developments in robotics and miniaturization have made it feasible to screen hundreds or thousands of conditions in parallel, accelerating the pace of research and discovery.

18.2.2 Advanced Analytical Techniques

Recent advances in analytical techniques, such as real-time monitoring of pH, dissolved oxygen, and biomass concentration, have enhanced the effectiveness of microtiter plate-based fermentations. Optical density measurements, fluorescence, and luminescence assays can provide insights into microbial growth and metabolism without disrupting the fermentation process. Moreover, integration with mass spectrometry and high-performance liquid chromatography (HPLC) allows for the analysis of metabolites in real time. This comprehensive monitoring capability enables researchers to gain deeper insights into microbial physiology, optimize growth conditions, and improve product yields.

18.2.3 Novel Microbial Strains and Genetic Constructs

The rise of synthetic biology has led to the development of novel microbial strains and genetic constructs optimized for specific bioprocesses. Microtiter plate-based fermentations serve as an ideal platform for testing and characterizing these engineered organisms.

Researchers can quickly assess the performance of genetically modified strains in various growth media and fermentation conditions. The ability to rapidly iterate on genetic designs, combined with high-throughput screening, has significantly accelerated the development of microbial cell factories capable of producing valuable biochemicals, biofuels, and pharmaceuticals. For example, engineered strains of Escherichia coli and yeast have been developed to enhance the production of biofuels and other high-value compounds, with microtiter plates serving as the initial screening platform.

18.3 IMPACTS ON SYNTHETIC BIOLOGY

18.3.1 Accelerated Discovery and Development

Microtiter plate-based fermentations have greatly accelerated the discovery and development of new microbial strains and bioprocesses. The ability to rapidly screen multiple conditions and strains has led to shorter development timelines and reduced costs. This is particularly important in the competitive field of synthetic biology, where time-to-market can significantly impact the viability of new products.

By enabling researchers to quickly identify optimal growth conditions and metabolic pathways, MTPs have facilitated the rapid development of microbial cell factories tailored for specific applications. As a result, innovations in bioprocesses, such as sustainable biofuel production and the generation of valuable metabolites, have become more accessible.

18.3.2 Enhanced Understanding of Microbial Physiology

The extensive data generated through microtiter plate-based fermentations has led to a deeper understanding of microbial physiology and metabolism. Real-time monitoring and high-throughput screening provide valuable insights into the relationships between environmental factors and microbial behavior. This understanding is crucial for synthetic biology applications, where precise control over metabolic pathways is essential for optimizing product yields. By elucidating the underlying mechanisms governing microbial growth and metabolism, researchers can make informed decisions when engineering organisms for specific biotechnological applications.

18.3.3 Facilitation of Metabolic Engineering

Microtiter plate-based fermentations play a significant role in metabolic engineering, allowing researchers to systematically explore and manipulate metabolic pathways.

By screening various genetic constructs and conditions, scientists can identify key enzymes and regulatory elements that influence metabolic flux. The ability to rapidly test multiple genetic combinations and fermentation conditions has enabled the development of engineered strains with enhanced production capabilities. This has broad implications for the production of biochemicals, pharmaceuticals, and biofuels, as researchers can efficiently optimize pathways to maximize yields.

18.4 FUTURE PROSPECTS

18.4.1 Integration with Omics Technologies

The future of microtiter plate-based fermentations lies in their integration with omics technologies, such as genomics, transcriptomics, proteomics, and metabolomics. Combining high-throughput fermentation with these techniques can provide comprehensive insights into microbial behavior and metabolic pathways. By analyzing the genomic and transcriptomic profiles of microorganisms in conjunction with fermentation data, researchers can identify key regulatory networks and metabolic bottlenecks.

This systems biology approach has the potential to revolutionize synthetic biology by enabling the design of highly efficient microbial cell factories.

18.4.2 Expansion of Applications

As microtiter plate technology continues to advance, its applications are expected to expand beyond traditional fermentation processes. Future developments may include the use of MTPs for screening microbial consortia, exploring microbial interactions, and developing new bioprocesses for the production of complex biomolecules. Additionally, the integration of microtiter plate-based fermentations with bioreactor systems may facilitate the transition from small-scale screening to large-scale production. This scalability will be essential for commercializing biotechnological innovations and meeting the growing demand for sustainable bioproducts.

18.4.3 Improved Sustainability

The shift towards sustainable biotechnology is becoming increasingly important, and microtiter plate-based fermentations can contribute to this goal. By enabling the efficient screening of microbial strains capable of utilizing renewable resources and minimizing waste, MTPs can facilitate the development of greener bioprocesses. Furthermore, the ability to optimize conditions for microbial growth and product formation will contribute to reducing the environmental footprint of bioprocesses. As researchers focus on developing sustainable solutions, microtiter plate technology will play a crucial role in addressing the challenges of climate change and resource scarcity.

18.5 CONCLUSION

Microtiter plate-based fermentations have emerged as a cornerstone of modern synthetic biology and bioprocess development, offering numerous advantages in terms of efficiency, scalability, and data generation. Recent advances in automation, analytical techniques, and the development of novel microbial strains have significantly enhanced the utility of MTPs in research and industry. The impacts of these advancements are far-reaching, accelerating discovery, enhancing understanding of microbial physiology, and facilitating metabolic engineering. As the field continues to evolve, the future of microtiter plate-based fermentations holds great promise, with potential expansions into new applications, integration with omics technologies, and contributions to sustainable biotechnology. Through these innovations, microtiter plate technology will continue to play a pivotal role in shaping the future of biotechnology.

CHAPTER 19

CONTROL AND MONITORING OF INDUSTRIAL FERMENTATIONS: AN INDUSTRIAL PERSPECTIVE AND RECENT TRENDS

Industrial fermentation is a critical process in the production of a wide array of products, including biofuels, pharmaceuticals, and food ingredients. As the demand for these products increases, so does the need for efficient, reliable, and scalable fermentation processes. Control and monitoring of industrial fermentations are essential components of this endeavor, ensuring optimal conditions for microbial growth and product formation.

19.1 THE IMPORTANCE OF CONTROL AND MONITORING IN INDUSTRIAL FERMENTATIONS

Effective control and monitoring of industrial fermentations are crucial for achieving consistent product quality, maximizing yields, and minimizing costs. Fermentation processes involve complex biochemical reactions influenced by various factors, including temperature, pH, dissolved oxygen, and nutrient availability. Any fluctuations in these parameters can adversely affect microbial activity and product formation, leading to suboptimal yields or even process failure.

19.1.1 Ensuring Optimal Conditions

One of the primary goals of fermentation control is to maintain optimal conditions for microbial growth and metabolism.

This involves real-time monitoring of key parameters such as temperature, pH, and dissolved oxygen levels. For instance, certain microorganisms thrive at specific temperatures, and deviations from this range can hinder growth rates and alter metabolic pathways. Automated control systems enable real-time adjustments to these parameters, ensuring that the fermentation environment remains conducive to microbial activity. By employing feedback loops and predictive algorithms, modern control systems can proactively address deviations, maintaining optimal conditions throughout the fermentation process.

19.1.2 Enhancing Product Quality and Consistency

In industries such as pharmaceuticals and food production, product quality is of utmost importance. Variability in fermentation conditions can lead to inconsistencies in product quality, affecting both safety and efficacy. By implementing robust monitoring and control systems, manufacturers can reduce this variability, ensuring that each batch meets stringent quality standards. For example, in the production of antibiotics, maintaining precise control over fermentation parameters is essential to achieve the desired purity and potency of the final product. Automated monitoring systems can track and adjust conditions in real time, facilitating consistent production quality across batches.

19.1.3 Reducing Operational Costs

Effective control and monitoring not only enhance product quality but also contribute to cost savings. By optimizing fermentation conditions, manufacturers can improve yields, reduce resource consumption, and minimize waste. This is particularly important in large-scale operations, where even minor improvements in efficiency can result in significant cost reductions. Moreover, real-time monitoring allows for early detection of issues, enabling timely interventions that can prevent costly downtime or product loss. By streamlining operations and reducing the need for manual oversight, automated control systems can enhance overall productivity and profitability.

19.2 RECENT TRENDS IN FERMENTATION CONTROL AND MONITORING

19.2.1 Automation and Digitalization

The integration of automation and digital technologies into fermentation processes has transformed control and monitoring practices. Automated systems allow for real-time data collection and analysis, enabling manufacturers to respond quickly to changes in fermentation conditions. This shift towards digitalization is characterized by the adoption of advanced sensors, process control software, and data analytics platforms. With the advent of Industry 4.0, fermentation processes are increasingly becoming part of interconnected systems that facilitate data exchange and collaborative decision-making. This connectivity enhances the ability to monitor processes remotely, optimize operations, and improve overall efficiency.

19.2.2 Advanced Sensor Technologies

Recent advancements in sensor technologies have significantly improved the ability to monitor fermentation processes. Next-generation sensors can provide high-resolution, real-time data on critical parameters such as pH, temperature, dissolved oxygen, and substrate concentrations. These sensors are often more accurate and reliable than traditional methods, enabling more precise control of fermentation conditions. For instance, optical sensors can monitor dissolved oxygen levels without the need for chemical reagents, reducing the complexity of monitoring systems. Similarly, advances in miniaturization and integration allow for the development of multi-parameter sensors capable of measuring multiple variables simultaneously.

19.2.3 Machine Learning and Artificial Intelligence

The application of machine learning (ML) and artificial intelligence (AI) in fermentation control and monitoring is gaining traction. These technologies enable the analysis of large datasets generated during fermentation processes, facilitating the identification of patterns and correlations that may not be evident through traditional analysis. By leveraging ML algorithms, manufacturers can develop predictive models that anticipate changes in fermentation conditions and product outcomes. These models can enhance decision-making processes, enabling proactive adjustments to maintain optimal conditions and improve product yields.

19.2.4 Real-Time Data Analytics

Real-time data analytics has become an integral part of fermentation control systems. The ability to process and analyze data in real time allows for immediate feedback and adjustments, enhancing the overall efficiency of the fermentation process. Modern control systems can visualize data through dashboards and alerts, making it easier for operators to identify trends and anomalies. The integration of cloud-based platforms enables remote access to fermentation data, allowing for collaboration among team members and facilitating more informed decision-making. This capability is particularly beneficial in multi-site operations, where centralized monitoring and control can optimize performance across various facilities.

19.3 CHALLENGES AND FUTURE DIRECTIONS

While significant advancements have been made in the control and monitoring of industrial fermentations, several challenges remain. One of the primary challenges is the integration of diverse data sources into a cohesive monitoring system. As fermentation processes become increasingly complex, the ability to synthesize information from various sensors and data sources will be critical.

19.3.1 Data Management and Integration

The proliferation of data generated during fermentation processes necessitates robust data management strategies. Manufacturers must develop systems that can effectively handle large volumes of data, ensuring that relevant information is readily accessible and actionable. This may involve the use of data lakes, advanced analytics, and machine learning algorithms to extract valuable insights from the data generated.

19.3.2 Addressing Variability in Fermentation Processes

Variability in fermentation processes remains a significant challenge, as even small fluctuations can lead to significant changes in product yields and quality. Future advancements in monitoring and control will need to focus on developing more sophisticated models that account for this variability. This may involve integrating real-time monitoring with predictive modeling to identify potential issues before they impact the fermentation process.

19.3.3 Sustainability Considerations

As industries move towards more sustainable practices, the control and monitoring of fermentation processes must also evolve. Manufacturers will need to implement strategies that minimize resource consumption, reduce waste, and lower environmental impacts. This may involve optimizing fermentation conditions to improve yields, recycling by-products, and adopting renewable resources as feedstocks. Innovations in monitoring systems can play a crucial role in achieving sustainability goals by providing real-time insights into resource utilization and environmental impacts. By enhancing efficiency and minimizing waste, manufacturers can contribute to a more sustainable bioprocessing landscape.

19.4 CONCLUSION

The control and monitoring of industrial fermentations are critical components of successful bioprocess development, impacting product quality, operational efficiency, and cost-effectiveness. Recent trends, including automation, advanced sensor technologies, machine learning, and real-time data analytics, have transformed fermentation practices, enabling manufacturers to optimize conditions and improve yields.

Despite the advancements, challenges remain, particularly in data integration, addressing variability, and promoting sustainability. As the industry continues to evolve, embracing innovative technologies and strategies will be essential for maintaining competitive advantages and meeting the growing demand for bioproducts. By prioritizing effective control and monitoring practices, manufacturers can ensure the success of industrial fermentation processes and contribute to the development of sustainable biotechnological solutions.

BIBLIOGRAPHY

Advanced Biotechnology By Kumar, D.S. and Kumar, D. IK International Publishing, 2011.

Advances in Fermented Foods and Beverages: Improving Quality, Technologies and Health Benefits Edited by Holzapfel, W. Woodhead Publishing, 2014.

Applied Dairy Microbiology Edited by Marth, E.H. and Steele, J.L. Marcel Dekker, 2001.

Biochemical Engineering and Biotechnology By Ghasem, N.M. and Henda, R. Elsevier, 2015.

Bioprocess Engineering Principles By Doran, Pauline M. Academic Press, 1995.

Bioprocess Engineering: Basic Concepts By Shuler, M.L. and Kargi, F. Prentice Hall, 2002.

Biotechnology for Beginners By Ratledge, C. and Kristiansen, B. Elsevier, 2006.

Biotechnology in the Chemical Industry Edited by Rehm, H.J. and Reed, G. VCH, 1986.

Biotechnology of Industrial Microorganisms By Demain, A.L. and Solomon, N.A. Elsevier, 1985.

Biotechnology of Lactic Acid Bacteria: Novel Applications Edited by Mozzi, F., Raya, R.R., and Vignolo, G.M. Wiley-Blackwell, 2010.

Biotechnology of Yeasts and Filamentous Fungi Edited by Chen, W., Druzhinina, I.S., and Kubicek, C.P. Springer, 2017.

Biotechnology: A Textbook of Industrial Microbiology By Crueger, W. and Crueger, A. Sinauer Associates, 1989.

Essentials of Industrial Microbiology By Lydon, J. CRC Press, 2015.

Fermentation and Biochemical Engineering Edited by Murray Moo-Young Springer, 2014.

Fermentation and Biochemical Engineering Handbook: Principles, Process Design, and Equipment Edited by Vogel, H.C. and Todaro, C.L. William Andrew Publishing, 1996.

Fermentation Microbiology and Biotechnology (2nd Edition) Edited by El-Mansi, E.M.T. and Bryce, C.F.A. CRC Press, 2002.

Fermentation Processes: Emerging and Conventional Technologies Edited by Ray, R.C. and Didier, M. CRC Press, 2021.

Fermented Foods in Health and Disease Prevention Edited by Frias, J., Martinez-Villaluenga, C., and Peña, G. Academic Press, 2016.

Fermented Foods, Part I: Biochemistry and Technology Edited by Ray, R.C. and Didier, M. CRC Press, 2016.

Fermented Milk and Dairy Products Edited by Tamime, A.Y. Wiley-Blackwell, 2013.

Food Fermentation: Microorganisms, Applications, and Production Edited by Hutkins, R.W. ASM Press, 2006.

Food Microbiology and Laboratory Practice By Sezer, D., Erol, O.B. and Erol, G.Y. Wiley-Blackwell, 2018.

Food Microbiology: Fundamentals and Frontiers Edited by Doyle, M.P. and Buchanan, R.L. ASM Press, 2012.

Fundamentals of Biotechnology By Chmiel, H. Springer, 1991.

Fundamentals of Industrial Microbiology By Patel, S. Springer, 2013.

Fundamentals of Microbiology and Food Fermentations By Adams, M.R. and Moss, M.O. Springer, 2007.

Handbook of Brewing: Processes, Technology, Markets Edited by Eßlinger, H.M. Wiley-VCH, 2009.

Handbook of Enology: The Microbiology of Wine and Vinifications By Ribereau-Gayon, P., Dubourdieu, D., Donèche, B., and Lonvaud, A. Wiley, 2006.

Handbook of Food and Beverage Fermentation Technology Edited by Hui, Y.H., Meunier-Goddik, L., Hansen, A.S., Josephsen, J., Nip, W.-K., and Stanfield, P.S. Marcel Dekker, 2004.

Handbook of Microbial Fermentation Edited by Winkler, S. Nova Science Publishers, 2010.

Handbook of Plant-Based Fermented Food and Beverage Technology Edited by Hui, Y.H. and Evranuz, E.Ö. CRC Press, 2012.

Industrial Fermentation Processes and Products By Jones, R. Plenum Press, 2001.

Industrial Microbiology and Biotechnology (3rd Edition) Edited by Baltz, R.H., Demain, A.L., and Davies, J.E. ASM Press, 2010.

Industrial Microbiology By Casida, L.E. Wiley, 1968.

Industrial Microbiology: An Introduction By Waites, M.J., Morgan, N.L., Rockey, J.S., and Higton, G. Wiley-Blackwell, 2001.

Microbial Biotechnology: Microorganisms in Industrial Processes By Gupta, P. and Roy, S.K. New Age International, 2009.

Microbial Biotechnology: Principles and Applications By Glazer, A.N. and Nikaido, H. Cambridge University Press, 2007.

Microbial Fermentation and Enzyme Technology Edited by Wang, H. and Demain, A.L. Elsevier, 1977.

Microbial Fermentation in Food Processing By Joshi, V.K. and Pandey, A. IK International, 2007.

Microbial Functional Foods and Nutraceuticals Edited by Liong, M.T. CRC Press, 2010.

Microbial Technology: Microbial Processes Edited by Peppler, H.J. and Perlman, D. Academic Press, 1979.

Microbiology and Biochemistry of Cheese and Fermented Milk Edited by Fox, P.F. Springer, 1993.

Microbiology and Biotechnology Edited by Mansi, E.M.T. El and Bryce, C.F.A. Taylor & Francis, 2007.

Microbiology and Fermentation of Traditional Foods By Ray, R.C. and Didier, M. CRC Press, 2014.

Microbiology of Fermented Foods Edited by Wood, B.J.B. Blackie Academic & Professional, 1998.

Modern Industrial Microbiology and Biotechnology By Okafor, Nduka CRC Press, 2007.

Principles of Fermentation Technology By Stanbury, P.F., Whitaker, A., and Hall, S.J. Butterworth-Heinemann, 2016.

Probiotic and Prebiotic Foods: Technology, Stability, and Benefits to the Human Body By Todorov, S.D. and Franco, B.D.G. Springer, 2020.

Wine Microbiology: Practical Applications and Procedures By Fugelsang, K.C. and Edwards, C.G. Springer, 2007.

Yeasts in Biotechnology and Human Health Edited by Querol, A. and Fleet, G.H. Springer, 2006.

INDEX

Z

www.ingramcontent.com/pod-product-compliance
Lightning Source LLC
LaVergne TN
LVHW062116200926
848484LV00007B/1683